AN INTRODUCTION
TO TTCN-3

AN INTRODUCTION TO TTCN-3

SECOND EDITION

Colin Willcock and Thomas Deiß

Nokia Siemens Networks GmbH & Co. KG, Germany

Stephan Tobies

European Microsoft Innovation Center, Germany

Stefan Keil

Research In Motion Deutschland GmbH, Germany

Federico Engler

TeliaSonera CIS, Sweden

Stephan Schulz

Conformiq Inc., Finland

A John Wiley and Sons, Ltd., Publication

This edition first published 2011
© 2011 John Wiley & Sons Ltd.

LTE is a trademark of ETSI. LTE and LTE Advanced logos have been reproduced by permission of
ETSI – http://www.3GPP.org/

Registered office
John Wiley & Sons Ltd, The Atrium, Southern Gate, Chichester, West Sussex, PO19 8SQ, United Kingdom

For details of our global editorial offices, for customer services and for information about how to apply for permission to
reuse the copyright material in this book please see our website at www.wiley.com.

Library of Congress Cataloging-in-Publication Data

An introduction to TTCN-3 / Colin Willcock ... [et al.].
   p. cm.
 Includes bibliographical references and index.
 ISBN 978-0-470-66306-6 (cloth)
 1. Telecommunication systems–Testing–Data processing. 2. Computer
networks–Testing–Data processing. 3. Programming languages (Electronic
computers) I. Willcock, Colin. II. Title.
 TK5102.84.I58 2011
 005.13–dc22

                                                        2010037017

Print ISBN: 978-0-470-66306-6 (HB)
ePDF ISBN: 978-0-470-97791-0
oBook ISBN: 978-0-470-97790-3
ePub ISBN: 978-0-470-97789-7

Typeset by Laserwords Private Limited, Chennai, India
Printed in Malaysia by Ho Printing (M) Sdn Bhd

Contents

List of Figures

List of Tables

About the Authors

Colin Willcock

Colin Willcock is currently manager for 3GPP Radio Access Network Standardisation at Nokia Siemens Networks. He received a BSc from Sheffield University in 1986, an MSc from Edinburgh University in 1987, and a PhD in parallel computation from the University of Kent in 1992. Colin was part of the core ETSI team that developed the TTCN-3 language and spent many years leading and participating in the TTCN-3 language maintenance. In the past, he has worked on numerous standardisation efforts at ETSI, ITU-T, and 3GPP, focusing on various aspects of formal specification languages. He was the project leader for the European TT-medal project, which strove to improve test methodology and languages for software-intensive systems and also lead the D-MINT project, which aimed to improve test methodology and languages for software-intensive systems and explored the use of model-based testing in an industrial context.

Thomas Deiß

Thomas Deiß is a Senior System Specialist at Nokia Siemens Networks. He received an MSc in Computer Science and a PhD in Natural Sciences from the University of Kaiserslautern in 1990 and 1999. He is currently specifying transport features for mobile communication systems. Before joining Nokia Siemens Networks, Thomas developed the Nokia Research Center TTCN-2 and TTCN-3-based test systems, developed course materials and taught courses about TTCN-3, and has participated for several years in TTCN-3 standardisation. He was a contributor to the European TT-medal and D-MINT projects, which strove to improve test methodology and languages for software-intensive systems and explored the use of model-based testing in an industrial context.

Stephan Tobies

Stephan Tobies is a Software Design Engineer at the European Microsoft Innovation Center where he works on software verification. He received an MSc in Computer Science and a PhD in Natural Sciences from the University of Technology in Aachen in 1998 and 2001. He has been actively involved with TTCN-3 until 2005 while working as a Senior Research Engineer at Nokia Research Center. During that time, he has been a member of ETSI Strategic Task Force 253, which was responsible for the maintenance and extension of the TTCN-3 standard. He has been a lead developer of an industry-grade TTCN-3 tool and has been working in the area of TTCN-3 language development and test system implementation.

Stefan Keil

Stefan Keil is a software developer at Research In Motion. He received an MSc in Electrical Engineering from the Ruhr University in Bochum in 1996. Stefan has worked for Alcatel as a programmer in the field of fixed line communications and a technical trainer for broadband communication fibre technology. From 2000 to 2007 he worked as a Research Engineer at Nokia Research Center in the area of test system implementation, TTCN-3 tool development, and training. At Nokia Siemens Network he worked from 2007 to 2009 in software specification for base station software. In 2009 Stefan started working for Research In Motion in the field of embedded software development on end user devices.

Federico Engler

Until December 2004, Federico Engler has been a Principal Engineer at Nokia Research Center. He studied computer science at Uppsala University from 1989 till 1993. After that, he started working for Telelogic, where he was involved in standardisation issues around TTCN-2, TTCN-3, and ASN.1, as well as TTCN-3 tool development. In January 2003, Federico started working for Nokia in the area of automated test solutions, which involved the mapping, documentation, and synchronisation of test-related activities at a Nokia-wide level. He has also been involved in activities around improved visualisation and documentation of tests and test results within Nokia. Federico is currently working for TeliaSonera CIS where he leads the development of portal-based telecommunications applications.

Stephan Schulz

Stephan Schulz is currently the Chief Technology Officer at Conformiq Inc. He received an MSc and PhD in Computer Engineering from University of Arizona at Tucson in 1997 and 2001. Prior to his positions at Conformiq he has worked as a resident testing expert at ETSI's Centre for Testing and Interoperability as well as a Senior Research Engineer at Nokia Research Center. Throughout his career he has been consulting different users and organisations on TTCN-3 deployment as well as test suite and test system development. He has been an editor of the TTCN-3 Runtime Interface (TRI) standard, lead TTCN-3 architect in various ETSI Specialist Task Forces, designer of ETSI's official TTCN-3 web site, and author of many publications on the testing of text-based protocols with TTCN-3. He has been developing and teaching TTCN-3 courses as well as co-chaired four TTCN-3 User Conferences. In 2010, he was elected chairman of ETSI's Technical Committee Methods for Testing and Specification which is overseeing TTCN-3 standardisation.

Foreword

TTCN first saw the light of day as a fledgling language in the mid-1980s. With a major modernization of TTCN over 10 years ago, resulting in TTCN-3, it has arguably progressed to be the *de facto* `international` standardised language for writing test specifications for reactive systems. Here at ETSI, TTCN is the cornerstone of many complex test specifications, covering a wide range of technologies, including 3GPP LTE™,[1] Intelligent Transport Systems, eHealth, Voice Over IP and IPv6.

I had the pleasure of working with Dr. Colin Willcock and Professor Jens Grabowski to produce the very first edition of the core specification of TTCN-3. Since that time, the language has gone from strength to strength. It has a growing body of users, good tool support and, most importantly, a dedicated and very active maintenance team. Indeed, TTCN-3 is a living language, continuous improvements and the addition of new features, demanded by the user community, means that this book too has been updated. This second edition addresses those new features admirably.

As is common with any programming language, the language specification is often not the first place a user will go to learn her new craft. Often, this will be done by reading a good textbook (or at least looking at the examples). Unfortunately, the TTCN-3 community has not had this luxury – until now, that is. This very first TTCN-3 book fulfils a long-awaited need and I see its publication as a milestone in the evolution of the language. The authors of this book are uniquely qualified to explain the details of TTCN-3 as applied to practical, real-life situations. They have a wide experience of contributing to the development of TTCN-3, building TTCN-3 tools and test systems and using the language in serious commercial projects, ranging from mobile communications to the automotive industry.

The essence of TTCN-3 is really quite simple, which partly explains its increasing popularity. The early chapters of this book capture this essence and provide the novice reader with a clear and intuitive introduction to the language. For the more demanding reader, subsequent chapters delve into the language in greater depth.

This excellent book is likely to be regarded as the definitive TTCN-3 user's companion for many years to come.

Anthony Wiles
European Telecommunications Standards Institute (ETSI)

---

[1] ™LTE is a trade mark of ETSI.

Preface

At this point in time, nearly 10 years after the first TTCN-3 core language standard was published and 5 years after the first version of this book appeared it seems the right moment to bring out a new revised and extended version. In this second version of the book we have integrated the 4000+ change requests that have been added to the language since the first version of the book came out. In addition we have added a new chapter on testing frameworks and a new chapter explaining LTE [1] testing using TTCN-3.

Looking back over the 10 years, I feel somewhat like a parent to the language. I was there all those years ago at its inception and I look back to those early days with nostalgia and affection. It was a pleasure to work in that small focused team at ETSI trying to define and develop a global testing language. Like any child we had high hopes for TTCN-3, but also great uncertainty of whether it would ever match those hopes and aspirations. After the first standards were published, there followed a hectic period of dissemination and one important milestone following another. The official TTCN-3 launch event, the first commercial TTCN-3 tool set, the TT-Medal European project, the first TTCN-3 book, the further development and maintenance of the language at ETSI. I had the pleasure of leading or involvement in all these development steps along the way.

Now, 10 years later the TTCN-3 language has grown, both in terms of use and functionality to become a global testing language in terms of geography and a wide spread testing technology in terms of industrial domains where it is used. TTCN-3 and automated software testing are no longer the major focus for me, with others taking over the mantel of maintaining and extending the language. To use the parent analogy, I feel the child is now making its own way in the world without me. However it is still with a certain pride and affection that I watch the success of the language from the side lines.

Colin Willcock
Nokia Siemens Networks (NSN)

Acknowledgements

We would like to thank those people without whom this book would never have been written. First and foremost are our families and friends, who supported us and bore our (physical or mental) absence while we wrote this book. Next we would like to thank the people in the ETSI task forces who have shaped TTCN-3 and continue to advance the language and its use in many areas. We would also like to thank the European ITEA programme, which has supported the transfer of TTCN-3 technology to European industry through the TT-Medal project. Lastly, we would like to thank the test engineers in our companies with whom we have worked and whose application of TTCN-3 in the real world has contributed much to our TTCN-3 experience.

Abbreviations and Acronyms

The text in this book contains a number of abbreviations and acronyms. The list below describes all their meanings in one single place.

ASN.1   Abstract Syntax Notation One
CORBA   Common Object Request Broker Architecture
DNS   Domain Name System
EMM   Evolved Mobility Management
EPS   Evolved Packet System
ETSI   European Telecommunication Standardisation Institute
FTP   File Transfer Protocol
GSM   Global System for Mobile Communications
IDL   Interface Definition Language
IP   Internet Protocol
ISO   International Standardisation Organisation
ISP   Internet Service Provider
IMS   IP Multimedia Subsystem
IUT   Implementation Under Test
ITU-T   International Telecommunication Union – Telecommunication
               Standardisation Sector
HTTP   Hyper Text Transfer Protocol
LTE   Long Term Evolution
MAC   Medium Access Control
MSC   Message Sequence Chart
MTC   Main Test Component
NAS   Non-Access Stratum
OMA BCAST Open Mobile Alliance Mobile Broadcast Services Enabler Suite
OMG   Object Management Group
PA   Platform Adapter
PTC   Parallel Test Component
RAT   Radio Access Technology
RFC   Request For Comments

RLC	Radio Link Control
RPC	Remote Procedure Call
RRC	Radio Resource Control
RTS	RunTime System
SA	SUT Adapter
SIP	Session Initiation Protocol
SMTP	Simple Mail Transfer Protocol
STF	Specialist Task Force
SUT	System Under Test
TCI	TTCN-3 Control Interface
TCP	Transmission Control Protocol
TE	TTCN-3 Executable
TETRA	Terrestrial Trunked Radio
TRI	TTCN-3 Runtime Interface
TSI	Test System Interface
TTCN-3	Testing and Test Control Notation Version 3
UDP	User Datagram Protocol
UE	LTE User Equipment or mobile device
UTRA	Universal Terrestrial Radio Access
VHDL	Very High Speed Integrated Circuit Hardware Description Language
WiMAX	Worldwide Interoperability for Microwave Access
XML	Extensible Markup Language
XSD	XML Schema

1

Introduction

The Testing and Test Control Notation Version 3 (TTCN-3) is an internationally standardised language for defining test specifications for a wide range of computer and telecommunication systems. It allows the concise description of test behaviour by unambiguously defining the meaning of a test case pass or fail. The predecessor of TTCN-3, TTCN-2, has been used successfully for over a decade, mostly in testing telecommunications systems. In this third revision of TTCN, the best parts of the previous testing language have been combined and extended with a powerful new textual syntax to create a universal testing language whose application area is no longer restricted to testing telecommunication systems.

Five years after the publication of the first edition of this book a lot of things have happened in the TTCN-3 community. TTCN-3 is celebrating its 10th anniversary and has grown into a 10 part standard with four extension packages to date. It has grown into a *global* testing language used well beyond telecommunication and standardisation. We are witnessing a rapid uptake of the language in Asia – especially India and China – which today provides already more than 40% of all testing services worldwide. TTCN-3 has established itself in the automotive and medical domain, with the banking sector promising to be the next big application area. The continued success and propagation of the language is visible in the annual international TTCN-3 User Conference which reflects these developments. In addition, TTCN-3 has further strengthened its position in standardisation and is used increasingly for certification and acceptance testing: In telecommunication, the third Generation Partnership Program (3GPP) [2] has decided to perform all of their future test suite development including their prestigious test suite for Long Term Evolution (LTE[TM1]) [1] /4G terminals using TTCN-3 (see Chapter 16). The Open Mobile Alliance (OMA) [3] is using TTCN-3 for their test suite development in service provision and broadcasting across mobile networks. The WiMAX Forum [4] is using TTCN-3 for certifying the conformance of terminals and the interoperability of terminals and network elements. In the automotive domain, the influential AUTOSAR consortium [5] – composed of all the major car manufacturers, suppliers and service

---

[1] [TM]LTE is a trade mark of ETSI.

An Introduction to TTCN-3, Second Edition.
Colin Willcock, Thomas Deiß, Stephan Tobies, Stefan Keil, Federico Engler and Stephan Schulz.
© 2011 John Wiley & Sons, Ltd. Published 2011 by John Wiley & Sons, Ltd.

providers – has adopted TTCN-3 for the specification of its compliance test suites. And in the internet domain, TTCN-3 test suites are used in the context of certification of the IPv6 ready program [6]. Last but not least the European Telecommunication Standardisation Institute (ETSI) is continuing to use TTCN-3 in a wide range of application areas including Next Generation Networks (NGNs), electronic passport, radio technology as well as evaluating test execution traces at their interoperability events [7].

This book provides a solid introduction to the TTCN-3 language and its use. All the important concepts and constructs of the language are explained in a tutorial style, with the emphasis on extensive examples. This book also introduces the larger picture of how the testing language is related to the overall task of test system implementation. By doing so, it becomes the perfect companion to the available TTCN-3 language standards [8–21], filling the gaps like style guide, structuring and application. In addition, this book points out some of the dangers and pitfalls of TTCN-3 on the basis of our personal TTCN-3 experience from language standardisation, tool implementation and applying TTCN-3 for a number of years *in the real world*. The style and level of this book make it suitable for both engineers, learning and applying the language in the real world, and students, learning TTCN-3 as part of their studies. Although this book is intended to be accessible to a wide audience, it does assume that the reader has some basic knowledge of software programming.

This second edition of the book has been updated and revised to cover the additions, changes and extensions to the TTCN-3 language since the first version was published. In addition to this extensive new material caused by language evolution the book also contains a new section on testing frameworks and a major new example domain: LTE testing using TTCN-3. This domain is not just covered in the text but also in the form of downloadable tools and a test suite. This book is structured to present concepts in an order that offers the quickest start to the efficient use of the TTCN-3 language. In Sections 1.1 and 1.2, we discuss the advantages of using TTCN-3. Chapter 2 then goes on to introduce a complete example to get a first hands-on impression of the language and lists all the additional parts that are necessary to transform the TTCN-3 code into a working test system. In Chapter 3, we then move on to present the basic language concepts of TTCN-3, including basic types, operators and expressions, as well as the language constructs for control flow. In Chapter 4, the subject of test specification in TTCN-3 is considered in more detail by discussing the language constructs that are most commonly used for non-concurrent testing. Test cases, test verdicts and message-based communication are a few of the topics that are considered. Concurrent TTCN-3 is described in Chapter 5; the key issues considered are the usage and synchronisation aspects of test components. The importance of procedure-based communication is highlighted by providing a separate chapter, Chapter 6, which provides an in-depth discussion of this communication paradigm. Chapter 7 considers the issue of modularity, which is needed to address the issues of code re-usability as well as multi-user development. Chapter 8 provides a thorough introduction to the TTCN-3 type system, leaving more complex type topics such as external type systems to Chapter 9. Templates and advanced aspects of their use are brought up in Chapters 10 and 11. Chapter 12 considers TTCN-3 extension packages in general and the extension package for real-time testing specifically. With all the language parts described in detail, Chapter 13 then provides a detailed description of how TTCN-3 test systems work in practice. In Chapter 14 we consider frameworks for testing and Chapter 15 can be seen as a utility chapter that provides a collection of code examples and common sense advice. Chapter 16 rounds off the book by introducing LTE testing using TTCN-3. This provides the link to the tools and test suite available on the companion website which will enable you not just to read about TTCN-3, but actually experiment with it.

1.1   TTCN-3 as a Language

TTCN-3 is a language designed specifically for testing. Many constructs are similar to those in other programming languages but are extended with additional concepts not available elsewhere. These concepts include built-in data matching, distributed test system architecture and concurrent execution of test components. TTCN-3 has a larger type system than normal programming languages and includes native types for lists, test verdicts and test system components. In addition, TTCN-3 provides direct support for timers as well as for message-based and procedure-based communication mechanisms.

TTCN-3 is an internationally standardised test language [8–21]. Within these documents, the meaning of each and every language element is clearly and precisely specified. This means that a test script written in TTCN-3 is unambiguous. This precise definition of the language also leads to tool vendor independence, since every tool should execute a given test case in exactly the same way. Tool vendor independence facilitates easy moving from one TTCN-3 toolset to another and greatly helps in testing projects where test tools from different vendors are used in parallel. The language is designed to provide a single general-purpose testing language suitable for a wide range of testing applications. It can be used across the whole product development cycle. In this way, TTCN-3 can provide major benefits in terms of return on investment in testing tools, training and, naturally, product quality.

At its heart, TTCN-3 has a powerful, intuitive textual format for defining test scenarios, that is similar to conventional procedural programming languages. This textual format is referred to as the *TTCN-3 core notation* [8]. This book concentrates on this core notation, with the following chapters describing in detail its syntax and use.

In addition to the core notation, TTCN-3 also supports the specification of test scenarios using other presentation formats. A TTCN-3 *presentation format* provides an alternative way of specifying test scenarios visually or in a context-specific manner. All presentation formats can be converted into the core notation while preserving their meaning as shown in Figure 1.1. Two presentation formats have been standardised initially. The standardised tabular presentation format [9] was designed to give the test developers the 'look and feel' of the existing TTCN-2 tabular format. This format was introduced to provide an easy migration path for existing TTCN-2 users into the TTCN-3 world. An example is shown in Figure 1.2. The graphical presentation format [10] uses an extended version of an MSC-like Message Sequence Charts [22] notation for specifying test scenario behaviour as shown in Figure 1.3. Neither of the two presentation formats has gained acceptance by the TTCN-3 community. Therefore they are not considered further in this book.

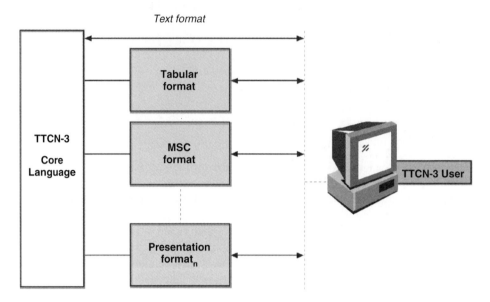

Figure 1.1   TTCN-3 presentation formats.

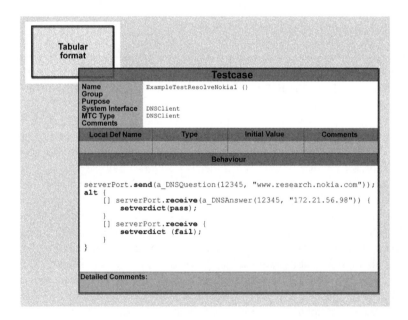

Figure 1.2   An example of the tabular presentation format.

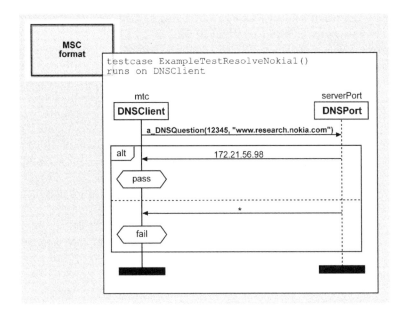

Figure 1.3   An example of the graphical presentation format.

1.2   The Development of TTCN-3

The direct predecessor of TTCN-3, TTCN-2, was developed by ISO as part of the overall methodology for testing protocol layers in the Open Systems Interconnection (OSI) seven-layer architecture [23]. TTCN-2 was first standardised in the late 1980s by ITU-T [24] and ISO [25]. It has been successfully applied within the area of conformance testing for telecommunications protocols. Nevertheless, a number of problems and shortcomings were limiting its possibilities to be used as a more general purpose testing language. In 1998, ETSI, the European Telecommunication Standardisation Institute, set up the Specialist Task Force (STF) 133 to develop a new improved version of TTCN, taking the known issues into account. This action resulted in the birth of TTCN-3. Over the following two years, the TTCN-3 language was developed with the involvement and input of most of the major tools and telecommunication companies. The official launch of the TTCN-3 language took place in October 2000 at Sophia Antipolis, France.

When developing TTCN-3, four major areas of improvement needed to be considered and addressed in relation to TTCN-2. These areas were productivity, expressive power, flexibility and extensibility.

The aspect of productivity was simply addressed by developing the core language to resemble other well-known, modern programming languages. By making TTCN-3 a textual language, it made it easier for users to edit and learn the new concepts. TTCN-3 also provides significantly extended functionality that makes the language powerful and suitable for a wider range of testing applications. Some of these extensions include better support of new types of testing such as special constructs and features for the testing of IP-based systems and text-based protocols like Session Initiation Protocol SIP [26].

Another major extension provides support for testing systems based on remote procedure calls, using, for example CORBA or web services.

Finally, TTCN-3 is extensible: TTCN-3 has explicit hooks and mechanisms built-in to the language that allow new features and notations to be easily integrated. Some new features are self-contained, for example the integration of IDL [27] and XML [28] type definitions as well as the definition of a common set of documentation tags. New parts in the set of TTCN-3 standards have been defined for these features, see [14–16]. Other new features are more multi-faceted and require the extension of the notation, the operational semantics, as well as other parts of the standard. Behaviour types, type parameterisation, as well as test deployment support are examples of such multi-faceted extensions. These extensions have been defined by separate *extension packages*, including the relevant modifications to the core language, operational semantics and the other parts of the TTCN-3 standards, see [17–19].

Both the introduction of new parts to the standard as well as the definition of extension packages, allow new features to be introduced while keeping the core language stable.

1.2.1   Future Development

TTCN-3 was designed to be a general-purpose testing language that could be used in many application areas. With time, this has spread its usage into many new fields where standardised testing languages have not been used before. Using TTCN-3 in these new areas has generated requests for additions to the language that can provide better support for testing requirements from domains such as real-time testing or performance testing.

TTCN-3 is actively maintained through a well-defined change request process handled by ETSI. The change request process provides a mechanism to balance the needs for stability and backwards compatibility with the calls for extended functionality from new users. Additionally, changes to TTCN-3 are well-documented and the reasoning behind the changes can be followed.

1.3   Summary

In this introduction, we have briefly presented the background and most important concepts of TTCN-3. We have seen that the language has a long history, which originates from the world of telecommunications. In the past decade, the worlds of telecommunications and the Internet have moved much closer together and the systems to be tested are constantly becoming more dynamic and complex in their nature. To meet these new challenges, the existing standardised test language, TTCN-2, was re-designed and extended to result in TTCN-3. The following chapter introduces an example that, even though simple in nature, contains many of the testing issues found in modern communication systems.

2

TTCN-3 by Example

To properly introduce the most important concepts of TTCN-3, we will start by looking at a real-life example. The example is based on the Internet's Domain Name System (DNS) [29] and aims at verifying that a DNS server is able to properly resolve host names to their corresponding IP addresses. The example in this section is highly simplified in order to allow us to focus on TTCN-3 related issues rather than on details of the particular problem domain.

When referring to the implementation or element, that is to be tested, the term *implementation under test* (IUT) is often used. If the IUT is part of a larger system and we can only communicate indirectly with the IUT and it is more appropriate to use the term *system under test* (SUT). In this book, we will only use the term SUT, as this naming is more general and applies also for the minimal case where we can talk directly with the IUT, and the IUT is the same as the SUT.

In the following sections, we will present an initial test case, showing how to test that a specific host name is correctly resolved. For this particular example, we will go through the necessary definitions for data types, messages and test behaviour. We will then extend this test case in three directions. Firstly, we will extend it to handle situations in which the SUT does not behave as expected. Secondly, we will show how several different interfaces of the SUT can be connected to different components of the test system to allow us to test different parts concurrently. The third and last extension will show how to use procedure-based communication as a potential complement to message-based communication.

2.1 TTCN-3 Test Suite

2.1.1 Problem Domain

To create a good test solution from scratch, a test developer needs to understand the problem domain. It is important to know the details about the information or messages that are going to be exchanged, the interfaces that are going to be used, and of course the

An Introduction to TTCN-3, Second Edition.
Colin Willcock, Thomas Deiß, Stephan Tobies, Stefan Keil, Federico Engler and Stephan Schulz.
© 2011 John Wiley & Sons, Ltd. Published 2011 by John Wiley & Sons, Ltd.

particular behaviour that needs to be verified. For this purpose, this section introduces the relevant aspects of the problem domain so that we can design a test solution that properly reflects the relevant parts.

The Internet's DNS is basically a large distributed database implemented by a world-wide collection of so-called *name servers*. These name servers contain information that allows applications to look up, in other words, resolve, the IP address for a given host. The need for such a system is rooted in the difference between how humans and machines handle information. Humans are good at handling images and simple names, but poor at handling numerical data. Machines are the other way around. To bridge the gap between these two worlds, the DNS hides IP addresses from humans and applications by allowing them to refer to hosts using mnemonic names, such as "www.nokia.com". Applications that need the IP number to perform a given action (for example Internet browsers with HTTP [30], mail programs with SMTP [31] or file transfer applications with FTP [32]) can connect to a name server to obtain the IP address on the basis of the name they have been provided with. This IP address is then used to connect to the remote machine.

Despite the enormous task of simultaneously resolving host names to IP addresses, the DNS works (mostly) reliably and fast because of the beauty of its design. Every owner of a subnet (for example Internet service provider (ISP) or company) is responsible for maintaining two or more local name servers that know the IP addresses of all the hosts within their own domain. If a user or application queries a local name server for the IP address of a host within the same domain, the server will immediately be able to provide an answer without the involvement of external DNS servers. The communication between the client and the local name server for this simple case is depicted in Figure 2.1.

If a query involves resolving the name of a machine in an external domain, the local DNS server will not immediately know the answer. Instead, it needs to turn to a so-called *root name server* to get information about yet another DNS server that probably knows the answer to the query. This more complex communication is depicted in Figure 2.2.

Even though the extra steps are not visible to the client, this situation needs to be handled differently by the local name server. This approach clearly keeps the information distributed in several places and for this reason requires a well-defined behaviour from its individual components. Our first test case will now have a look at how we can test the correct behaviour of a local name server that knows the IP address for a host within its

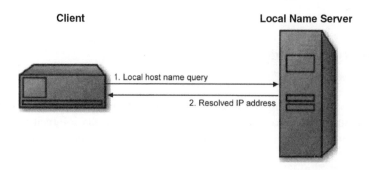

Figure 2.1   The few steps needed to resolve a local host name.

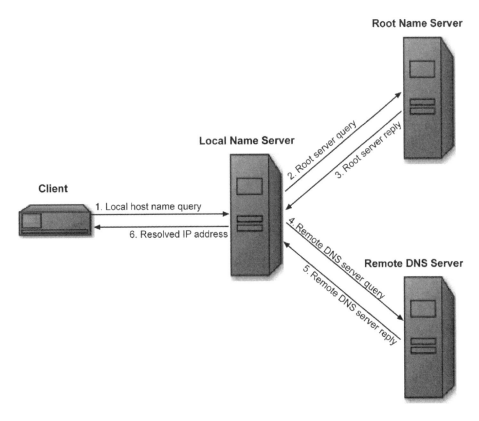

Figure 2.2 The steps of resolving a remote host name.

own domain. We will, in this section, gradually extend the coverage of our test definitions in order to allow us to deal with more and more complex situations.

2.1.2 Test Purpose

Understanding the problem domain is one of the major prerequisites for creating good tests. Documenting and understanding which parts of the problem domain need to be tested is an equally important requirement. A test purpose is a description that describes in prose or in some more formal manner (for example message sequence charts (MSCs)) the objectives of a given test. Test purposes can be used both as documentation for individual tests and as guidance for test writers that need to implement tests on the basis of some kind of description. The choice of notation for test purposes is subject to ongoing discussions, but as guidance, it can be useful to remember that the more informal a test purpose description is, the greater is the risk that the description can be interpreted in different ways by different people. Ambiguous descriptions are clearly something that should be avoided at all costs.

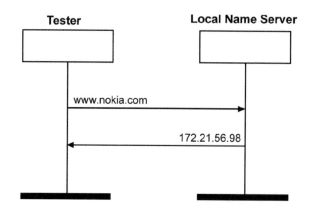

Figure 2.3   A simple test purpose described with an MSC diagram.

In this chapter, we will be using simple MSCs like the one in Figure 2.3 to describe the purposes of our tests. From this diagram, we can deduce that the test only involves the tester and the local name server. We can also deduce that the test is performed by first sending a query for "www.nokia.com" to the local name server, which should then be followed by the reception of the correct IP address, which is 172.21.56.98. It is of course possible to add more information to this MSC diagram and also to combine it with prose to remove as much ambiguity as possible from the test description.

The problem we are trying to solve in our example is to test a local name server to make sure it is able to correctly resolve both local and remote host names. The reason for this is obvious. If our local name server is unable to function correctly, the communication between machines in the subnet will break down completely, or communication will be directed to unexpected partners. This is a serious situation that needs to be avoided by properly testing the server before deployment in the network.

2.1.3   TTCN-3 Modules

Before we start developing our test cases, we need to know how we should structure our test code. TTCN-3 code is collected in TTCN-3 modules. A module can contain test definitions and also a control part that defines how the different tests are to be executed. A module can import definitions from other modules, providing a flexible mechanism for modularisation. To be able to start defining our data types, values and test cases, we first need to create a module. Table 2.1 shows how this looks in TTCN-3 code. From the figure you can also see that TTCN-3 comments may use both C and C++ style, that is a comment is either enclosed with "/*" and "*/" or it reaches from a "//" to the end of the line.

2.1.4   Data Types and Messages

Before we can define any test cases in our test module, we need to have a look at the messages that are going to be exchanged between the test system and the SUT. The

Table 2.1 Creating our first TTCN-3 module

```
/* -------------------------------------------------------------
 * File: DNSTester.ttcn
 * Desc: This is our small test suite for testing some simple name
 *       server behaviour.
 * -------------------------------------------------------------
 */

module DNSTester {
  // Here we will add our definitions.
  // Here we will add our control part.

  // The control part must come after the definitions.
}
```

communication between the client and the DNS server takes place by using DNS messages [29]. Real DNS messages are rather complex, and we will not be going into the details of them in this book. What we will do is to simplify such a message to a format that better fits our simple example. In general, it is good to remember that the structure of the messages to be exchanged is not arbitrary. The structure of these messages is often defined by a standard or other document that describes in detail the implementation we are about to test.

For our purposes, there are only two types of DNS messages. These are question and answer messages. Furthermore, these messages have the same format so a particular flag in the message is needed to indicate if the message is a question message or an answer message. A DNS message also contains an identification field, that is a 16-bit integer generated by the client application in order to allow the client to identify the answer for a previously generated query. Real DNS messages also contain a body part that allows a single DNS message to contain several questions or several answers simultaneously, but in our example we limit the number of questions or answers in a single DNS message to one.

Our DNS messages will be defined using the TTCN-3 record type. Records are used to define ordered structured types that are collections of basic or other structured type elements. In Table 2.2, our DNS message type is defined as a record that contains the four elements that make up our simplified messages. To represent the message kind field, we have used an enumerated type to create an enumeration with the two elements e_Question and e_Answer. To improve the readability of our code, we use prefixes for identifiers to give an immediate hint of what the identifier represents. Chapter 13 explains in further detail the naming conventions used throughout this book. The message identification is represented with an integer field that has been subtyped to the range of values that can be represented by 16 bits. Our question and answer fields are represented as character strings. The answer element in our messages is marked optional as it will not be present in DNS question messages.

Once we have defined the types we are going to use, we need to start thinking about the actual instances of the types – the messages – that are going to be exchanged in the test process. In TTCN-3, these 'instances' are called *templates*. Templates are used to either transmit specific values or to test whether received values are contained in the set of

Table 2.2   The definition of our DNS message type

```
// Simple type definitions to match the protocol structure

type integer      Identification( 0..65535 );              // 16-bit integer
type enumerated   MessageKind {e_Question,e_Answer};
type charstring   Question;
type charstring   Answer;

// The definition of our DNS message type.

type record DNSMessage {
  Identification  identification,
  MessageKind     messageKind,
  Question        question,
  Answer          answer           optional
}
```

expected messages, which are represented by a template specification. In our example, we will be using templates for sending queries to the SUT and for matching incoming replies.

Templates are powerful because they allow not only specific values to be specified, but also ranges, lists and matching attributes that together provide a compact and powerful mechanism to describe *sets* of expected messages and provide automatic checking that received data conforms to the specifications we have described. Templates will be handled in more detail later on in this book. Here, we just introduce them briefly.

A template for a given type must specify a value or matching expression for each and every field of its type. Table 2.3 shows the definition of a template for our DNS message type. In this case, the template represents a DNS question as the messageKind field is set to e_Question. The identification field is set to an arbitrary number, in this case 12345, and the question refers to the host name to be resolved: "www.nokia.com". As a question must not contain an answer part, the answer field has been marked absent with the attribute omit.

There is a drawback with our definition in Table 2.3 and, that is that the template is rather hard-wired. When we want to send several questions to our DNS server by defining them in this manner, it would mean that we would have to define a template for each individual question, and each of these definitions would contain the same messageKind

Table 2.3   A send template for our DNS message type

```
// A possible template for the DNS message type.

template DNSMessage a_NokiaQuestion := {
  identification    := 12345,
  messageKind       := e_Question,
  question          := "www.nokia.com",
  answer            := omit
}
```

Table 2.4   A parameterised send template for our DNS questions

```
// A parameterized template for DNS questions based on DNSMessage.

template DNSMessage a_DNSQuestion( Identification p_id,
                                   Question p_question ) := {
  identification    := p_id,
  messageType       := e_Question,
  question          := p_question,
  answer            := omit
}
```

Table 2.5   A parameterised receive template for our DNS answers

```
// A parameterized template for DNS answers based on DNSMessage.

template DNSMessage a_DNSAnswer( Identification p_id, Answer p_answer ) := {
  identification    := p_id,
  messageType       := e_Answer,
  question          := ?,
  answer            := p_answer
}
```

and answer field. To avoid these repeated definitions, we can use parameterisation to allow a more flexible and reusable solution.

Table 2.4 shows a modification of a_NokiaQuestion. The template has been renamed to a_DNSQuestion as it now can be reused for all DNS questions. Two parameters have been added for those fields that change between different questions. The first parameter is the identification number that needs to be different for each individual question to allow correct mapping to incoming answers. The second parameter contains the string with the actual host name we want to resolve. As a DNS question always has the messageKind field set to e_Question, this particular field does not need to be parameterised. This also applies to the answer field that always is omitted in question messages.

We now also need to define a template for the expected DNS answers. Table 2.5 defines a parameterised template, which is similar to a_DNSQuestion in Table 2.4. It contains two parameters. The first one is for the identification number and the second is for the string that represents the expected reply from the server. The messageKind field is fixed as replies always have this field set to e_Answer and the question field is marked with the matching attribute ?, which means that this field may contain any value, but that we ignore what this value is. We can do this because the identification field contains the same id as the outgoing question, and this allows us to couple them more effectively.

2.1.5   Components and Ports

With our data types and templates ready to be used, we now need to start looking at the test solution from an architectural point of view. TTCN-3 allows a test developer to

Table 2.6 The definition of our single port and single component

```
// DNS messages are allowed to move in and out through ports of this type.

type port DNSPort message {
  inout DNSMessage
}

// Our single component uses one single port to communicate with the SUT.

type component DNSClient {
  port DNSPort serverPort
}
```

use a single or several test components to perform a testing task. The components can communicate with each other and with the SUT. The points at which communication takes place are called *ports*. A port is modelled as an infinite first-in-first-out (FIFO) queue in the receive direction. The queue stores incoming messages or calls until they are processed by the component that owns that particular port. Our initial example will only use one single component and one single port to communicate with the SUT. Each port has a type, which defines the used communication paradigm (message- or procedure-based communication) and specifies the types of messages that can be sent and received by that port. Table 2.6 shows how we can define a port type in TTCN-3. Our port type is called DNSPort and uses message-based communication. The example specifies a port type that can both send and receive messages of type DNSMessage. It is also possible within a port type specification to only send messages of a certain type (out) or to only receive messages of a certain type (in). In our example, the same type of DNS message is used for questions and answers, so our port type allows DNS messages in both directions.

The single component for the initial example will only have a single port of type DNSPort. The component is named DNSTester and contains one single port, which we name serverPort, as shown in Table 2.6.

2.1.6 A First Test Case

We are now finally ready to start writing our first test case. This test case will be very simple; indeed, it will be incapable of dealing with an erroneous SUT. For the purpose of providing a simple example, we will assume that we are performing the test within the nokia.com domain and that we are looking for the IP address of www.research.nokia.com. From our test system, we will query the local name server for this IP address and observe what actually happens.

Table 2.7 shows how small our initial test case turns out to be with the definitions we have given so far. The test case is called tc_ExampleTestResolveNokia1 and runs on a DNSTester component. The test case initiates execution by sending a DNS question via the serverPort port to the SUT asking for the IP address of www.research.nokia.com. It then waits for a matching incoming answer that should contain the IP address 172.21.56.98. In case such a message is received, the test case sets the verdict to pass and stops execution. TTCN-3 allows you to set different verdicts

Table 2.7 Our first test case assumes the correct answer will arrive without problems

```
// Our first test case! This small test case will behave very poorly in case
// of an erroneous SUT. More about this later!

testcase tc_ExampleTestResolveNokia1() runs on DNSClient {
  serverPort.send( a_DNSQuestion( 12345, "www.research.nokia.com" ) );
  serverPort.receive( a_DNSAnswer( 12345, "172.21.56.98" ) );
  setverdict( pass );
  stop;
}

// Our small control part.

control {
  execute( tc_ExampleTestResolveNokia1() );
}
```

on the basis of the results of a given test. The `pass` verdict is used to specify that a given test has passed and the SUT has behaved as expected. The `fail` verdict is used to specify that a given test has not passed because the SUT behaviour was not as expected. The `inconc` verdict is used when it is impossible to deduce whether the observed behaviour is a `pass` or a `fail`, for example because the test system was unable to communicate with the SUT.

Observe finally that the test expects the incoming answer to contain the same identification number that was sent out with the original question.

Now that we have defined our first test case, the last step to enable it to execute is to call it. This test case execution is defined within the control part of the module. Table 2.7 shows the control part that is needed to run this single test case.

Our first test case turned out to be very compact and elegant, but it has a couple of weaknesses, which we will have a look at in the following section.

2.1.7 Handling Erroneous Situations

The first problem with `tc_ExampleTestResolveNokia1` that we need to address is its inability to handle unexpected answers from the SUT. When a message arrives at the port of our test system, the message is checked – in TTCN-3 terms *matched* – to see if the incoming values conform to the template for the particular receive statement. In the current test case, if an incorrect identification number comes in or an IP address does not match what we are expecting, then the `receive` statement will block forever. The received message will stay on top of the input queue without ever being removed because there is no `receive` alternative that can match and remove an unexpected reply at this point in the test case.

To resolve this situation, we can use the TTCN-3 `alt` construct, which allows us to specify that several different alternatives of behaviour can take place at a given point. An `alt` statement that contains several alternatives will block until any one of its alternatives matches. If we extend our initial test case and add an alternative that will match incorrect replies, the test case will not block in the same manner, and we will be able to state that

Table 2.8   Our extended test case is now able to handle incorrect incoming replies

```
// Our modified test case is now able to properly handle incorrect/invalid
// incoming messages.

testcase tc_ExampleResolveNokia2() runs on DNSClient {
  serverPort.send( a_DNSQuestion( 12345, "www.research.nokia.com" ) );
  alt {
    // Handle the case when the expected answer comes in.
    [] serverPort.receive( a_DNSAnswer( 12345, "172.21.56.98" ) ) {
        setverdict( pass );
      }
    // Handle the case when unexpected answers come in.
    [] serverPort.receive {
        setverdict( fail );
      }
  }
  stop;
}
```

the incoming message was not the expected one and thus the test has failed. Table 2.8 shows the modified test case. After the send statement, we have added an alt statement that contains two alternatives. The first one is the same receive statement that we had in our original test case. The second one is a receive statement without parameters, which means that it matches any incoming message, which has not been previously matched by any of the alternatives above it.

The test case in Table 2.8 is able to handle incorrect replies, but it does not cover the case when a reply might never turn up. If the name server is down or seriously congested for some reason, the test case will still be blocked until a reply comes in, which might never happen. This situation needs to be handled in a better way and can be resolved by using timers. If a timer is started when the DNS question is sent out, we can specify that we require the incoming reply to show up within a given amount of time. By catching timeouts, we can now extend our test case further to handle the problem of missing replies, as shown in Table 2.9. A timer called replyTimer is started and set to run for 20 seconds directly after the DNS question is sent. The alt statement has been extended with a third alternative, that is able to catch a timeout from the timer if no reply is received from the SUT within this time.

2.1.8   Default Behaviour

When reading the previous section, you will probably have noticed that TTCN-3's way of dealing with unexpected or untimely SUT behaviour could lead to considerable code duplication: if, for every receive statement that can fail to match, we need to add at least two additional cases to catch incorrect or missing responses, then our test cases will soon grow out of proportion and become impossible to maintain or even understand. For this reason, TTCN-3 contains a construct called *default behaviour*. Default behaviour can be seen as a catch mechanism that allows the test author to handle unexpected situations implicitly. Instead of having to write TTCN-3 code to handle these situations explicitly

Table 2.9 Our test case is now able to also handle missing answers

```
// Our test case is now able to handle incorrect replies as well as
// missing replies.

testcase tc_ExampleResolveNokia3() runs on DNSClient {
  timer replyTimer;
  server.send( a_DNSQuestion( 12345, "www.research.nokia.com" ) );
  replyTimer.start( 20.0 );

  alt {
    // Handle the case when the expected answer comes in.
    [] serverPort.receive( a_DNSAnswer( 12345, "172.21.56.98" ) ) {
        setverdict( pass );
        replyTimer.stop;
      }
    // Handle the case when unexpected answers come in.
    [] serverPort.receive {
        setverdict( fail, "Unexpected response from SUT" );
        replyTimer.stop;
      }
    // Handle the case when no answer comes in.
    [] replyTimer.timeout {
        setverdict( fail, "No response since 20 sec from SUT" );
      }
  }
  stop;
}
```

in each place where they may occur, the test author can do this in one single place and define that such behaviour should be used implicitly when none of the explicitly available alternatives matches. Default behaviour will be handled in detail later on in this book.

2.1.9 Multi Component TTCN-3

So far in this section, we have simplified things to make sure we only use one test component and one port. In real-life testing, situations are often more complex than this, involving more than one interface of the SUT. In many cases, tests that require access to more than one interface can be adequately structured by having one dedicated test component per interface. In the following example, we are going to extend the tests of our name server so that we examine how the server behaves when it receives a query that it is not able to answer itself. If such a query reaches a local name server, the server consults a root name server to obtain the address of a third server, that is supposed to know the answer (in the real world, some caching of recently resolved names is kept within the local name server, but for the sake of our example, we assume that such a cache does not exist and hence every non-local query leads to a request to a root name server). Root name servers differ from local name servers in that they maintain extensive lists of domains and of name servers responsible for those domains. The root name server will in most cases not provide the final answer itself (root servers are few in the world and need to avoid being congested), but will provide the address to an authoritative name server for the domain that contains the host we are trying to look up. This means that our local name server needs to first communicate with a root name server and then with at least one more, remote DNS server to obtain a reply for the initial query. These steps have been previously depicted in Figure 2.2.

If we take a closer look at our local name server, then it is connected to the Internet via its network link, meaning one single point of physical connection. Logically, on the other hand, the name server can be seen as having two kinds of interfaces. The first interface is the interface used by clients or applications to query the server (usually via User Datagram Protocol (UDP) port 53). The second interface is the network interface that the server uses when it itself needs to place a new query. These two interfaces are depicted in Figure 2.4.

It would be possible to extend our test case to test the scenario described above using a single test component, but this requires the test case to handle all possible permutations of message exchanges between the involved parts, and such a test case would explode in size and become difficult to understand and maintain. It is far better to create multiple test components that act as the client, the root name server, and the remote name server, respectively. We will not show or develop the TTCN-3 code for such a multi component solution this early in the book. We will rather just explain which steps need to be taken in order to create this test solution.

Figure 2.4 identifies that our SUT has two different interfaces. These interfaces need to be described at the TTCN-3 level in what is called the *test system interface* (TSI). The TSI defines the common interface that different test components will share towards the SUT when the tests are executed.

Once we have identified the TSI, we need to focus on the test components that are going to take part in our tests. For the scenario we wish to test, we are going to use four different test components. One component is called the *Main Test Component* (MTC) and is responsible for creating the parallel test components needed for the test (as well as to collect their individual verdicts and calculate a global, final verdict for the whole test). In our example, we have decided that the MTC does not participate actively in the tests itself, but there is no limitation in the language and you can decide to use a more 'active' MTC if you wish. Apart from the MTC, we are going to use three additional parallel test components. One of them will take the role of a client sending the query that the local

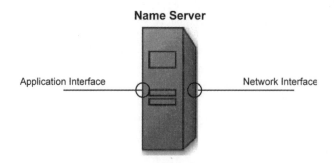

Figure 2.4 Logically, a name server has an application and a network interface.

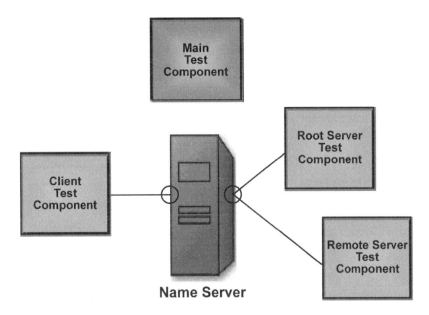

Figure 2.5 The configuration for our multi component test using four parallel test components.

name server is not able to answer on its own. This component will be connected to the application interface of the SUT and run basically the same behaviour that we used in the single component case. A second test component will act as the root server and the third component will take the role of the remote name server, which the root name server specifies as the authoritative server. These two last components will be connected to the network interface of the SUT.

By default, the MTC is the first component that executes. It will start off by creating the three parallel test components that we need for our test. Once these components are created, the MTC will map their ports to the TSI as depicted in Figure 2.5. When this configuration is established, the MTC will start different test behaviours on the different test components. The two test components on the network side are passive and wait for actions to take place from the SUT. The client component, on the other hand, is active and will initiate the whole test when it starts to run. Each parallel test component will reach its own, individual verdict reflecting its view of the test execution. The MTC will wait until all parallel test components have terminated, and then (automatically) calculate the final verdict.

2.1.10 Procedure-Based Communication

Until now, we have been looking at message-based communication. It is important to highlight that TTCN-3 can also handle procedure-based communication. To give an example of this, we will have a look at a different interface of the local name server that we can control and combine with the tests we have previously defined. As we have learned so far, a local name server keeps a table with mappings from host names to IP addresses

for its domain. Let us assume, for the purpose of our example, that the test system has
access to a management interface that allows it to control and manipulate the contents
of the name server table before or during the execution of tests. This would allow us to
control the presence or absence of entries in the mapping table and would provide us with
greater control and diversity over what we are testing.

In our example, we use procedure-based communication to manage the local name
server by using its available management software interface. We wish to make calls to this
interface to put the server in some defined initial state before the tests are initiated. To be
able to use procedure-based communication, we first need to define so-called *procedure
signatures*. These signatures give us information about the parameters that need to be
present in a call as well as information about return values or exceptions that might be
raised. We will use signatures to specify procedures in the SUT that can be called from
within the test system, but of course signatures can also be used the other way around,
meaning that the SUT can call procedures provided by the test system. In Table 2.10, we
see the signature definitions for three management procedures that allow us to clear the
mapping table, and to add and delete entries from the table. Following these definitions,
a procedure-based port type is declared so that only the test system can invoke these
procedures in 'outgoing' calls. If the keyword `in` had been used instead, it would have
meant the SUT was allowed to call the procedures in the test system.

To be able to call the procedures specified by these signatures, we now need to specify
signature templates in a similar way as is done for structured types in message-based
communication. Instead of referring to a type when creating these templates, we refer
to a signature instead. The elements in the template are then simply the parameters in
the signature. Table 2.11 shows how we can create a parameterised template for the
`AddEntry` signature.

Table 2.10   Signatures for procedures and the definition of a procedure-based port

```
// Our three signatures for the management of name table contents.
signature ClearTable () return boolean;
signature DeleteEntry( in charstring name ) return boolean;
signature AddEntry   ( in charstring name, in charstring ip_addr )
return boolean;

// Our procedure-based port for remote management of table contents.
type port ManagementPort procedure {
  out ClearTable, DeleteEntry, AddEntry
}
```

Table 2.11   A parameterised template for the `AddEntry` signature

```
// A parameterized template for the AddEntry procedure signature.
template AddEntry a_AddEntry( charstring p_name,
                             charstring p_ipAddress ) := {
  name       := p_name,
  ip_address := p_ipAddress
}
```

Table 2.12 Using procedure-based communication from inside a TTCN-3 function

```
function f_ClearMappingTable( ManagementPort p_mgmtPort ) return boolean {
  var boolean v_result;

  // Make a call and wait a maximum of 10 seconds for a reply.
  p_mgmtPort.call( a_ClearTable, 10.0 ) {
    // If the reply takes place, save the return value and then return it
    [] p_mgmtPort.getreply( a_ClearTable ) -> value v_result {
        return v_result;
    }
    // If no reply shows up and we get a timeout, return false.
    [] p_mgmtPort.catch( timeout ) {
        return false;
    }
  }
}
```

Now that we have our procedure signatures and signatures templates, we are able to start issuing calls to these procedures. Let us assume that our test cases, before starting to communicate with the SUT on its client interface, first make sure to clear the server's mapping table and then initialise it with a set of known values. Table 2.12 shows how a TTCN-3 function can be used to implement the part that clears the mapping table. The function has no parameters but returns a boolean value depending on whether its actions were performed successfully or not.

The function starts out by issuing the call to the ClearTable procedure, using the a_ClearTable signature template. Observe that it is possible and recommended to use a time limit for how long the test system will wait for a reply from the called procedure. The code block following the call – the call statement's body – is able to catch several different reactions to the procedure call. In the best of cases, the procedure returns and provides some kind of return value (if that has been specified in the signature). If for some reason no reply is returned, the code in Table 2.12 is able to catch a timeout and take following actions from that. It is also possible to specify in a signature definition if the called procedure can raise exceptions of different kinds. If the procedure can do that, the body of a call to such a procedure should also contain additional catch alternatives to handle such exceptions.

In this example for procedure-based communication, we have only introduced how the test system can issue calls to recipients outside the test system code. To make this test example even more interesting, it would also have been possible to keep the mapping table code inside the test system and allow the SUT to issue look-up calls that the tester itself would have been able to answer in either correct or incorrect ways. Even if such an example would have been interesting to show here, it is simply too complex to include in this tutorial book.

2.2 TTCN-3 Test Systems

So far, we have introduced TTCN-3 code segments that together make up a collection of simple tests. This collection is often referred to as a *test suite* and in this particular case we refer to it as an *abstract test suite*. The reason it is abstract is because it lacks any system-specific information, like how messages need to be encoded or how communication with the SUT actually takes place. In our simple examples, where we send and receive messages, we are never talking about the details how these messages are sent in the real world. We are not mentioning anything about the bits or bytes that are going to be exchanged between test system and SUT, or via which media the transmission is going to take place. This abstraction is valuable when creating a test suite as it removes system-specific details from the description of the test case behaviour, but to end up with real-life tests that execute and interact with the real SUT, we need to move from the abstract world into the concrete one.

Like any other programming language, TTCN-3 code is not executable by itself. It either needs to be interpreted or translated into some executable format. Additionally, we have to add information that allows the tests to execute against the real SUT. The following parts outside of the TTCN-3 code need to be provided.

- **Codecs.** The messages defined in our tests need to be encoded into some format, that is understood by the SUT before they are sent. Conversely, received messages will be decoded from their encoded form into TTCN-3 value representation.
- **SUT adaptation.** All our message exchanges in the abstract test suite are defined as operations referring to a specific port. When we use a TTCN-3 construction like `serverPort.send( a_template )`, we do not specify what `serverPort` actually represents in the real world. The mapping of what a TTCN-3 port actually represent in the real world, and the mapping between TTCN-3's communication mechanism and that of the SUT, need to be done in the SUT adapter.
- **Platform adaptation.** To handle situations when messages go missing, we have introduced the use of timers in some of our example test cases. As timers are implemented differently on different platforms and in different testing scenarios we sometimes require different notions of time, the actual timer implementation needs to be provided by the test system developer. The calling of TTCN-3 external functions is also platform specific and hence also needs to be provided by the test system developer.
- **Test management.** TTCN-3 provides the test developer with a control part to specify the order in which the tests in the test suite should be executed. This is an acceptable approach for stable test environments, where the tests and their order seldom change. This approach, however, is less acceptable for test systems where it is important to be able to constantly introduce changes in test execution without the need of time-consuming re-compilations. Test management can provide better support for the creation of test campaigns or for the customisation of log formats and log handling.

To make sure that the previously listed functionality is added in an ordered and well-defined manner, there exist two additional standard documents that describe the interfaces that need to be used for this purpose. The first interface of interest is the TTCN-3 Runtime Interface (TRI) [12]. The TRI defines the operations for the SUT and platform adapters, respectively. The second interface is the TTCN-3 Control Interface (TCI) [13]. The TCI focuses on issues around test management, logging, encoders and decoders.

2.2.1   High-Level View of a Test System

Figure 2.6 shows a high-level view that summarises the anatomy of a test system. For a detailed description of this subject, please refer to Chapter 13. In this section, we will briefly introduce these different parts to give you a better feel for the overall mechanics of test execution.

The box labelled 'Generated Code' in the middle of the picture represents the behaviour specified on the TTCN-3 level, but in a suitable executable form. The module on its right represents the TTCN-3 runtime system, which implements the TTCN-3 operational semantics [11]. These two modules are often referred to as the *TTCN-3 Executable* (TE).

Below these two modules, the TRI interface specifies a set of functions that are used to allow abstract operational concepts such as communication and timers to be mapped to the specific SUT and execution environment.

The TCI interface consists of four sub-interfaces. The test management interface (TCI-TM) is used to control the creation and execution of tests. The coding/decoding interface (TCI-CD) is used to allow for the specification of external codecs. The

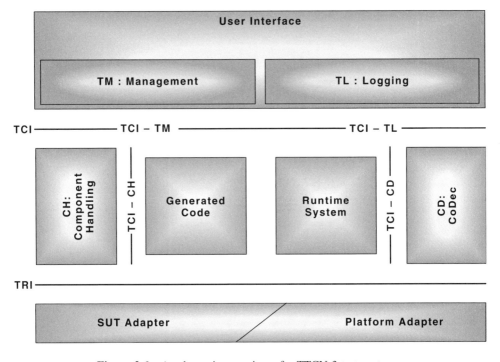

Figure 2.6   A schematic overview of a TTCN-3 test system.

component-handling interface (TCI-CH) allows the test system developer to specify how components are created and implemented when the test system is actually deployed. Finally, the logging interface (TCI-TL) can be used to create execution logs.

2.3   Summary

In this introductory chapter, we have highlighted TTCN-3's two major strengths, its unique and powerful testing concepts and its standardised interfaces. The advanced testing concepts in TTCN-3 such as ports for message- and procedure-based communication, timers, multiple test components, test verdicts and implicit templates matching provide an efficient and abstract way to specify the behaviour of the test systems. The standardised interfaces TCI and TRI allow adaptation of TTCN-3 test systems to virtually any kind of SUT.

3

Basic TTCN-3

In the first two chapters of this book, we have provided you with a high-level introduction of the most important aspects of TTCN-3 as a testing language. In this chapter, we will make the first transition to code level, gradually introducing language concepts that will lead to our first complete code example. It is expected that the reader has at least some prior programming experience with modern programming languages to fully benefit from the information in this chapter.

A TTCN-3 test suite is made up from one or more modules. A module is identified by a unique name and may contain a definitions part and optionally also a control part. The control part can be seen as similar to the main function in other programming languages and should only be present at one given place. The following sections will concentrate on the constructions found in the definitions part, which include among others, variables, constants, expressions and operators, as well as type and function definitions.

3.1 Basic Constructs

In this section, we will take a more detailed look at some of the language constructs that can be found in the module definitions part. We introduce identifiers and the rules that apply for their naming. We also explain the scoping and visibility rules of the language and we start off by having a look at constants and variables, leading on to data types, templates and eventually functions. The example code will be related to the DNS server domain, which has been introduced previously in the first two chapters.

3.1.1 Identifiers

Identifiers, like in any other programming language, are used to uniquely identify named entities in your code. TTCN-3 identifiers must consist of alphanumeric characters and may contain underscores. Identifiers must always start with a letter and are case-sensitive. This means that a variable named v_count is not the same as a variable named v_CoUnT. It

An Introduction to TTCN-3, Second Edition.
Colin Willcock, Thomas Deiß, Stephan Tobies, Stefan Keil, Federico Engler and Stephan Schulz.
© 2011 John Wiley & Sons, Ltd. Published 2011 by John Wiley & Sons, Ltd.

is recommended that a naming convention in the TTCN-3 code is used to improve both the understanding and maintainability of what you write. A few useful guidelines around this subject can be found in Chapter 15.

3.1.2 Modules

All TTCN-3 code must be specified within a module. A module is a top-level container for code that provides the user with the ability to improve reuse of given code segments.

A module is defined by using the keyword `module` followed by a unique name and a body within curly brackets, as shown in, for example Table 3.1. The module body can be empty, but consists often of at least a definitions part and an optional control part. The control part of a module is explicitly started with the use of the `control` keyword and specifies within curly brackets how the different test cases defined in the definitions part are to be executed once execution takes place.

3.1.3 Scope

Within TTCN-3, scope is conceptually defined by code blocks that are enclosed by curly brackets. Blocks of code can contain new individual code statements or new nested blocks. The outermost and top-level scope is the actual module.

Scoping in TTCN-3 is used, as in other programming languages, to control the visibility of particular language statements. Definitions within a particular code block are only visible within the code and nested scopes of that particular block. TTCN-3 forbids the

Table 3.1 The main structure of a TTCN-3 module

```
/* -------------------------------------------------------------
 * File: hostLookup.ttcn
 * Desc: This is our small test program to look up an IP-address
 *       in a local data structure
 * -------------------------------------------------------------
 */

module hostLookup {
  // The definitions part defines data structures and constants
  const integer c_maxTestNumber := 8;

  // The control part executes the dynamic part of the program.
  control {
    var charstring v_IP;
    var integer v_count          := 0;
    var integer v_Count          := c_maxTestNumber;
    // definition with constant reference and expression
    const integer c_halfTheTests := c_maxTestNumber / 2;
  }
}
```

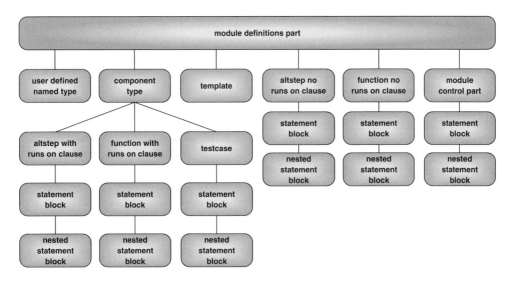

Figure 3.1 Scope hierarchy.

reuse of identifiers that occur in an outer scope. For example, in Table 3.1, reusing the identifier c_maxTestNumber in the control part is not allowed.

In reality, TTCN-3 scope units are more varied than just code blocks. Formally there are nine basic scope units; module definitions part, control part of a module, component types, functions, altsteps, test cases, statement blocks, templates and user defined named types. As some of the entities in this list have not yet been considered in this book and will only be described later, it is perhaps most helpful to just consider the scope hierarchy shown in Figure 3.1 to get the basic overview.

TTCN-3 requires identifiers to be declared before they can be referred to. The only exception to this rule are definitions on module scope, which can be made and referred to in an arbitrary order.

3.1.4 Constants

Constant definitions are denoted by the const keyword and can be placed at any given scope level. The value of the constant must be assigned at the point of declaration and is not allowed to change after this assignment. The programmer is allowed to use any arbitrary constant expression including pre-defined functions such as conversion functions to specify the assigned value. The random number generation function rnd() is not to be used in a constant definition. References to other constants within these expressions are allowed as long as these references are made without creating cycles.

The proper usage of constants is a must in the creation of understandable and maintainable code. Not only do named values provide a better understanding of what the code actually does but they also provide a single point of change in case a particular value needs to be altered. In Table 3.1, we use the integer constant c_maxTestNumber to define the number of host names that shall be resolved in our example. If the user decides to change the number of resolvable host names, it only needs to be changed at this single place. All the other places where this value is used, for example in the control part, do not have to be changed. Module parameters are another construct closely related to constants that can be used for this purpose which will be introduced (Chapter 7).

3.1.5 Variables

Variables in TTCN-3 work as in any other programming language and are used to save temporary values at run time during program execution. Variables are declared using the `var` keyword and can be defined at any scope level except at the top module level. The fact that variables cannot be declared at the module level means that TTCN-3 does not allow global variables. The reason for this limitation can be found in the problems that would otherwise occur when distributed test components would make changes to these variables and the updated value would need to be distributed to all the other concurrent test components in a given test.

In Table 3.1, the `integer` variable `v_count` is declared in the control part to count the number of host names resolved at run time within the program. It is possible to change its value at any time during execution using assignments.

When a variable is declared, it can be initialised with a value of the appropriate type, but this is not mandatory as in the case of constants. Reading a variable or using it in an expression before it is initialised results in a run-time error. In our example, `v_IP` is not assigned an initial value at its declaration. Therefore any attempt to use `v_IP` in an expression would cause a run-time error.

3.1.6 Comments

Readability and maintainability of source code generally improve if the author inserts proper comments. In TTCN-3 comments may contain any graphical character defined in ISO/IEC 10646 [33]. The language offers both line and block comments also known from other programming languages. A block comment starts with `/*`, can extend over several lines and ends with `*/`, while a line comment starts with `//` and extends to the end of the line – Table 3.1 contains examples for both kinds of comments. Documentation comments, a specific format that can be processed by external tools, are defined in part 10 of the standard [16], for more detail see Section 13.6.

3.1.7 Basic Data Types

TTCN-3 is a typed language with a large number of built-in types. In fact, the type system is so extensive that its detailed description requires its own dedicated chapter (see Chapter 8). In this section, we introduce only some simple data types and subtyping mechanisms. The motivation for this is to familiarise you first with the other powerful features of TTCN-3 to enable a quick start into writing real code.

The types we introduce here are `integer`, `boolean` and `charstring`. Apart from the types themselves, we will be using variables and constants that are bound to a specific type through their declaration.

Table 3.2 Examples of basic data types

```
module hostLookup {
  const integer c_maxTestNumber := 8;

  control {
    // variable declaration without initialization
    var integer v_count;
    // separate variable initialization
    v_count := 0;
    var boolean v_canBeResolved := false;
    // variable declaration with initialization
    var charstring v_IPAddress   := "134.23.16.157";
  }
}
```

Values of type integer can be positive or negative whole numbers, including zero. The modified example in Table 3.2 shows the declaration of the integer constant c_maxTestNumber, which is assigned the value 8. In the control part, the integer variable v_count is declared. In this case, the variable is initialised in the line after its declaration and can subsequently be used in expressions.

The type boolean consists of the two distinguished values true and false. Typically, a variable of type boolean is used to handle conditional operations. In Table 3.2, the variable v_canBeResolved is initialised to the value false. In the following program flow, the variable is used to check if a host name could be resolved successfully from the look-up table.

The type charstring represents a sequence of ASCII characters. Values of charstring are denoted by an arbitrary number of (printable) characters preceded and followed by double quotes. In Table 3.2, we have defined the charstring variable v_IPAddress representing the current IP address in the program flow. Note that, unlike in other programming languages, it is not possible to use escape sequences to express non-printable control characters, like the new line or the tab character. Section 8.3 explains how built-in functions resolve this problem.

3.1.8 Subtypes

At this point, we introduce two subtyping mechanisms, which are needed in Chapters 4 and 5. In many cases, it can be very useful if the allowed value range of a type can be restricted to a certain subset, thus creating a subtype. For example, a byte is an integer value restricted to non-negative values smaller than 256. Creating a subtype results in a new type-definition, which can then be used in the declarations of variables or constants.

The type integer and any other ordered type can be subtyped to a range of its values, by specifying an upper and lower bound for the allowed values. A new subtype is defined by using the keyword type, followed by the parent type, the name for the newly defined type, and the subtype's restriction. For example, the definition of the type

Table 3.3 Subtyping of basic data types

```
module hostLookup {
  type integer    Byte        ( 0 .. 255 );
  type charstring HostName ( "www.google.com", "www.leo.org",
                             "www.etsi.fr" );
  // example subtype use
  const Byte         c_userNo            := 4;
  const charstring c_IPaddresses[3] := {"127.0.0.1",
                                         "134.74.13.129",
                                         "209.42.14.134"};

  control {
    // host name from the value list
    var HostName  v_host := "www.google.com";
    // v_IP  == "134.74.13.129"
    var charstring v_IP   := c_IPaddresses[1];
  }
}
```

Byte in Table 3.3 shows the definition of such a subtype. Any constant or variable of such a subtype is required to conform to the subtype restrictions, and an assignment outside of the allowed values will cause an error, either during compilation or run time.

Another useful subtype restriction can be defined via a value list, that is a complete list of all legal values for the subtype. The values must be taken from the parent type (e.g. integer or charstring). For example, the charstring subtype HostName from Table 3.3 is restricted to a few Internet host names.

With the introduction of subtyping, type compatibility becomes an issue. Generally TTCN-3 requires type compatibility of values in assignments, instantiations, expressions and comparisons. A detailed explanation of type compatibility is given in Chapter 8. For the moment, it suffices to say that a variable can be assigned a value of another type as long as it is of the same root type and the value is within any associated subtype constraints of that variable.

For our example, we need a fixed list of host names and IP addresses. In this case, arrays can be used to create indexed lists of values of a type. The (positive, integral) size of the array must be given at the point of declaration between square brackets. Table 3.3 contains the definition of the constant array of charstring c_IPaddresses that contains three character string values. Each string can be accessed by specifying the name of the array and the index of the string. The control part in our example shows an example of such an access, that is the access of the second character string in the array c_IPaddresses. Arrays like this are indexed starting from 0 and any attempt to access a value outside of the permitted range will lead to an error.

3.1.9 Functions

Functions are generally used to structure a module by moving often-executed code or complex computations into separate, reusable elements. Functions can be called from the module control part, from test cases or other functions. When functions are called, they execute the statements in their body and then return execution to the point from where they have been called.

Functions are defined in the module definitions part with the `function` keyword, a unique name, a (possibly empty) parameter list, an optional return value and the function body. A function body may contain local constant and variable definitions and statements to express behaviour. As we have mentioned in Section 3.1.3, redefining identifiers that have already been used in outer scopes is not permitted.

Functions may specify a return value or template. In this case, the keyword `return` has to be specified after the parameter list in the function header followed by the return type. The function body must then contain at least one `return` statement, often the last statement, followed by a value or template, that is compatible with the specified type in the function header. When a `return` statement is reached inside the function body, the execution of the function is terminated and the specified value or template is returned to the calling context.

If the function header declares a parameter list, values can be passed into and out of the function at run time. Parameters are declared with an optional passing mode, their type and their name. By default, parameters are passed by value into a function, which means that it is possible to use constant values as actual parameters. Any changes to the parameter within the function will not be copied back when the function returns. The keyword `in` can optionally be used in front of the type in the parameter list to denote passing by value. The keywords `out` and `inout` *must* be used if parameters are to be passed by reference. Contrary to passing by value, a parameter passed by reference cannot be instantiated with a constant value. Changes to the parameter are copied back and remain visible when the function returns. The parameter list in the function header defines the exact sequence and types that must be used when instantiating the function. But it is also possible to provide actual parameters in arbitrary order by referring to the parameter name.

Table 3.4 defines two functions with input parameters and return values. The function `f_findHost` has one `in` parameter `p_host`. This parameter is designed to pass in the host name to be looked up. In the current version of the function, it always returns 0, because we haven't added any functionality to it yet. The `f_findHost` function is called from `f_resolve` and the result of the call is assigned to the variable `v_index`. The function `f_resolve` takes as its first parameter the host name to look up. The second parameter `p_ip` is used to pass back the resulting IP address. Note, to enable the value to be passed back from the function, `p_ip` is passed by reference. In addition to the `out` parameter, the function `f_resolve` returns a `boolean` value to indicate the success of the look-up operation. As the function `f_findHost` is not implemented, the return value is fixed to 0.

After the call to `f_resolve` in the control part, the charstring `v_IPAddress` contains the resolved IP address `"127.0.0.1"` for the host name `"localhost"` and `v_canBeResolved` evaluates to `true`.

Table 3.4 Basic use of functions

```
module hostLookup {
  const charstring c_hostNames[2]    := { "localhost", "www.nokia.com" };
  const charstring c_IPaddresses[2] := { "127.0.0.1", "207.34.94.128" };
  const charstring c_testNames[2]    := { "localhost", "local.host" };

  // should return the index of the matched host name in c_hostNames
  // or -1 if unmatched, currently not properly implemented
  function f_findHost ( in charstring p_host ) return integer {
    return 0;
  }

  function f_resolve( in charstring p_host, out charstring p_ip )
  return boolean {
    var integer v_index := f_findHost( p_host );
    // return true if the index is > = 0 or else false
    p_ip := c_IPaddresses[v_index];
    return true;
  }

  control {
    var charstring v_IPAddress;
    var boolean v_canBeResolved := f_resolve( c_testNames[0], v_IPAddress );
  }
}
```

Functions may also be defined externally by using the `external` keyword in front of the function prototype. The functional specification and realisation of external functions are outside the scope of pure TTCN-3. For more information about the invocation mechanism for external functions, refer to Chapter 12. The only constraint on external functions is that it is not allowed to perform any port operations within their behaviour.

3.1.10   Pre-Defined Functions

Next to user defined functions TTCN-3 also offers a number of very useful pre-defined functions. These include an extensive set of value conversion functions, for example convert an integer to a character, string handling functions, length and size functions, presence checking functions, codec functions, as well as some other special functions, for example for a random number generation function or checking if a template specifies a specific value. Rather than presenting these functions here we will introduce them as part of the constructs with which they can be used in the following chapters.

3.1.11   Parameters with Default Values

In real test systems there is often some test behaviour, that is used again and again in different test cases. Such test behaviour is often defined in libraries. The SUT may evolve over time, but often different releases of a SUT have to be tested. The test libraries have to evolve together with the SUT. A typical enhancement of a test library is to add new parameters to existing definitions. If all formal parameters have to be provided in all instantiations, then the addition of a new formal parameter requires that all instantiations of a parameter list are extended with a new actual parameter. If several releases of a test library are in simultaneous use, this would be quite a tedious and error prone activity, and without any actual benefit for the previous test systems. To ease the evolution of test libraries by adding parameters without breaking previous releases, TTCN-3 allows formal parameters with a default value to be defined. Parameters with default values can be left out in actual parameter lists.

Any **in** parameter in a parameter list in TTCN-3 may have a default value. Neither **out** nor **inout** parameters may have a default value. The default value is provided in the definition of the list of formal parameters. Obviously, the default value has to be compatible with the type of the parameter. The default value is used if no actual parameter is provided for a formal one. For example, if the trailing formal parameters in a parameter list have a default value, they can simply be left out from actual parameter lists. In the example in Table 3.5, the function f_findHost has one parameter with a default value. If no actual parameter is provided in a function invocation as in the control part shown, then the value "localhost" is used instead.

If there are other parameters following a parameter with a default value, then a parameter can be left out by using a dash instead of a value in the actual parameter list. In the example in Table 3.5 the first parameter of the invocation of f_resolve has been left out by using a dash instead.

Instead of providing actual parameters one by one as defined in the formal parameter list it is also possible to define explicitly which actual parameter belongs to which formal one. This is done by 'assigning' the actual parameters to the names of the formal parameters. This assignment notation can be used for **in, inout** and **out** parameters. It is neither permitted to mix the conventional and assignment notation nor to assign a parameter

Table 3.5 Use of parameters with default values

```
module hostLookupWithDefaultValues {
  function f_findHost ( in charstring p_host := "localhost" )
                        return integer {
    //some function body
  }

  function f_resolve( in charstring p_host := "localhost",
                      out charstring p_ip ) return boolean {
    //some function body
  }

  control {
    var charstring v_IPAddress;
    var boolean v_canBeResolved := f_resolve( -, v_IPAddress );
    f_findHost(); // same as f_findHost( "localhost");
    f_findHost( p_host := "www.nokia.com"); // assignment notation
  }
}
```

twice in an actual parameter list. If an actual parameter is not 'assigned' a value, then its default value is used. If an default value has been provided, then this is considered to be an error.

Although parameters with default values as well as the explicit assignment notation for parameters increase the flexibility in how to provide actual parameters, it is recommended to use these concepts with care and in a consistent way.

Note that in the TTCN-3 standards as well as in this book the term 'optional parameters' has not been used to avoid any confusion with optional fields of record and set types and to avoid the question of whether omit would denote an absent actual parameter.

Default values may be used in formal parameter list of almost all parameterized TTCN-3 constructs including (external) functions, test cases, altsteps, templates and parameterised types. Signature parameters may not have default values.

3.2   Basic Statements

This section introduces basic expressions and simple constructs like conditional statements and loops. Based on these we will create a first complete example that is actually computing something. In addition to conditional statements and loops, we will also introduce a number of behavioural statements and operations in this section, which will be extensively used in the following chapters.

3.2.1   Operators, Expressions and Assignments

Program flow is based on decisions that are often decided by calculations. Therefore, within TTCN-3 we need the ability to define expressions by combining data using operators. TTCN-3 has a number of built-in operators that allow the construction of complex expressions from literals, constants or variables. The operators are divided into the categories arithmetic (+, -, *, /, mod, rem), relational (==, <, >, != , >=, <=), logical (not and, or, xor), binary string (not4b, and4b, xor4b, or4b) and string (&, <<, >>, <@, @>). Operators are given a priority, which rules how complex expressions are interpreted (see Table 3.6).

In expressions with several operators, grouping of operands follows the priority rules. When necessary, a different grouping can be achieved using parentheses. Operations are evaluated from left to right, following the grouping established by the priority rules, where operators of higher priority have a stronger binding than those of a lower priority. In general, all operands used with arithmetic, logical or string concatenation operators must have the same root type.

Variables are updated by assignments using the := operation. During the execution of such an assignment, the right-hand side of the assignment must evaluate to a value, which is of a compatible type to the left-hand side. After the evaluation of an assignment statement, the variable stores the result of expression on the right-hand side.

An expression can combine compatible values using operators. Table 3.7 shows the assignment of a boolean expression to the variable v_isValidHostname. All variables used in expressions must be bound to a value, when the expression

Table 3.6 Priority of operators

Priority	Operator type	Operator
Highest		(...)
	Unary	-, +
	Binary	*, /, mod, rem
	Binary	+ , -, &
	Unary	not4b
	Binary	and4b
	Binary	xor4b
	Binary	or4b
	Binary	<<, >>, <@, @>
	Binary	<, >, <=, >=
	Binary	==, !=
	Unary	not
	Binary	and
	Binary	xor
	Binary	or
Lowest		

Table 3.7 Use of basic expressions

```
const charstring c_localDomain := "nokia.com";
// v_longName is set to true if a host name has too many chars between two
// dots, v_pairOfDots is set to true if a host name has two consecutive dots
var boolean v_longName := false, v_pairOfDots := true;
var boolean v_isValidHostname := ( ( not v_longName ) and
                                   ( not v_pairOfDots ) ) );
var charstring v_host        := "www" & "." & c_localDomain;
var integer v_a := 1, v_b := 2, v_average;
type integer    Byte      ( 0 .. 255 ); // integer subtype definition
var Byte v_c := 3;

// all variables must be initialized prior to the execution
// of this expression
v_average := ( v_a + v_b + v_c ) / 3;
// v_c has same root type as v_a and v_b
```

is evaluated. The charstring variable v_host is assigned to the result of the concatenation of the charstring literal "www", the literal "." and the charstring constant c_localDomain. The integer variable v_average is assigned the arithmetic average value of the values stored in v_a, v_b and v_c. Note that it is required that all variables are initialised prior to the calculation of the boolean expression or the average value. This is different to many other programming languages, where the use of uninitialised values in expressions will lead to unpredictable results. TTCN-3 guards against these kinds of errors and will catch such situations as run-time errors.

Table 3.8 Chained `if-else` statements

```
if ( ( v_count - v_lastDot ) > c_maxLabelLength + 1 ) {
    v_longName   := true;
    v_pairOfDots := false;
}
else if ( v_count  == v_lastDot + 1 ) {
    v_longName   := false;
    v_pairOfDots := true;
}
else {
    v_longName   := false;
    v_pairOfDots := false;
}
```

3.2.2 The Conditional Statements

For conditional execution in the program flow, the `if-else` statement can be used to branch on a boolean expression. Table 3.8 shows how the `boolean` variables `v_longName` and `v_pairOfDots`, which were used in Table 3.7, are set. The variable `v_longName` is set to `true` if the number of characters between two dots in the string is larger than `c_maxLabelLength`. If this is not the case but two consecutive dots are found, then `v_pairOfDots` is set to `true`. If neither is true, both variables are set to `false`. This example shows that it is possible to nest or chain `if-else` statements, basing one condition on another.

Similar as in other programming languages, TTCN-3 also supports a `select - case` statement to branch on conditions. If one or more values specified in a `case` branch of this statement matches the variable or expression referenced in its `select` header then the body of this branch is executed. The use of a `case else` branch provides the ability to unconditionally enter a branch. When no `case` branches match and no `case else` branch is specified in this statement, code execution continues with the next statement following the `select-case`. An example of such a construct is shown in Table 3.9.

3.2.3 Loops

Iterative or repetitive behaviour can be constructed with one of the three different loop constructs in TTCN-3: the `for` statement, the `do-while` statement or the `while` statement. In addition, TTCN-3 also offers `break` and `continue` statements to allow the exit and skipping of code in loop constructs.

The function `f_findHost` in Table 3.10 uses a `for` loop with the index variable `v_loop` as an iterator. The value of `v_loop` is increased by one after each loop as long as it is smaller than `c_hostNameCount`. The `for` statement header contains the definition of the variable `v_loop` and initialises it to 0. The exit condition of the loop is given as the ssecond of the three `for` statement parameters, which are divided by

Table 3.9 Example `select-case` statement

```
charstring v_char length(1);
... // assume that a value has been assigned to v_char
select ( v_char ) {
  case ( "A" .. "Z" ) {
    log( v_char, " is an uppercase character!" );
  } case ( "a" .. "z" ) {
    log( v_char, " is a lowercase character!" );
  } case ( "0" .. "9" ) {
    log( v_char, " is a numeric character!" );
  } case ( "1" ) { // dead branch due to
                   // previous case
    log( v_char, " is the number one!" );
  } else case {
    log( v_char, " is not an alphanumeric character!" );
  }
}
```

semicolons. The body of the loop statement will be repeated as long as the exit condition evaluates to `true`. The last statement in the header contains the expression, that is used to modify the iterator after each loop cycle. This modification occurs before the exit condition is evaluated. In our example, the loop is guaranteed to be terminated after at most c_hostNameCount iterations, in which case the function will return -1 to its caller. Inside the loop, the host name passed as parameter p_host into the function is compared to each of the host names in the look-up table. If the host name equals to an entry in the list, the index of this entry is returned by the function.

The function f_checkHostName uses a `while` loop to check the validity of the given host name before it is looked up in the table. The variable v_count counts up to the length of the `charstring` variable v_name. Once its value is equal to the length of the name string, the loop will terminate. This function also exemplifies the use of the `break` statement in loops. In this case it enables us to avoid the definition of a second termination criterion for exiting the `while` loop based on v_longName in case the string provided for the host name exceeds a length of 64 characters.

The function f_checkHostNameWithFor shows an alternative definition of the f_checkHostName function which exhibits the same behaviour but this time using a `for` loop and the `continue` statement. Here the `continue` statement skips the completion of the `for` loop iteration in case a character is a dot. After the execution of this statement the exit condition of the `for` loop is evaluated as the next step. Also the example shows that a loop can be terminated with a `return` statement. Contrary to the `break` statement this leads not only to the exit from the loop but also to the immediate return of the f_checkHostNameWithFor function.

A do-while loop is similar to a `while` loop with the difference that the loop exit condition is given after the loop, and is always checked at the end of each iteration, as demonstrated in Table 3.11.

Table 3.10 Functions with parameters and a return value

```
function f_findHost( in charstring p_host ) return HostIndex {
  for ( var integer v_loop := 0; v_loop < c_hostNameCount;
                    v_loop := v_loop + 1 ) {
    if ( c_hostNames[v_loop]  == p_host ) {
      return v_loop;
    }
  }
  return -1;
}


// check for names without dots or more than 63 characters between dots
// using while loop
function f_checkHostName( in charstring p_name ) return boolean {
  var integer v_lastDot  := 0;
  var integer v_count    := 0;
  var boolean v_longName := false;

  while ( v_count < lengthof( p_name )) {
    if ( p_name[v_count]  == "." ) {
      if ( ( v_count - v_lastDot ) > 63 ) {
        v_longName := true;
        break;
      }
      v_lastDot := v_count;
    }
    v_count := v_count + 1;
  } // end while
  if ( ( v_lastDot  == 0 ) or v_longName ) {
    return false;
  }
  return true;
}


// check for names without dots or more than 63 characters between dots
// using for loop
function f_checkHostNameWithFor( in charstring p_name ) return boolean {
  var integer v_dots   := 0;
  var integer v_count    := 0;

  for ( v_count := 0; v_count < lengthof( p_name ); v_count := v_count + 1 ) {
    if ( p_name[v_count]  == "." ) {
      v_dots := v_dots + 1;
      continue; // skip counting in case of a dot
    } else if ( v_count > 63 ) {
      return false;
    }
    v_count := v_count + 1;
  } // end while
  if ( v_dots == 0 ) {
    return false;
  }
  return true;
}
```

Table 3.11 `log`, `label` and `goto` statements by example

```
do {
  if ( f_checkHostName( c_testNames[v_count] ) ) {
    if ( f_resolve( c_testNames[v_count], v_ip ) ) {
      log( "host name has been resolved successfully" );
    }
    else {
      log( "host name '", c_testNames[v_count], "' could not be resolved!");
    }
  }
  else {
    log("'" , c_testNames[v_count], "' is an invalid host name!");
    // use of goto is possible, but discouraged: use of break would be
    // preferred
    goto exitLoop;
  }
  v_count := v_count + 1;
} while ( v_count < c_maxTestNumber );

label exitLoop;
```

3.2.4 Labels and Goto

TTCN-3 has the `label` and `goto` mechanism to provide the possibility to jump from one part of the program to another part. These have mainly been introduced because they were common in TTCN-2 and are useful for the conversion of existing test cases to TTCN-3. Their use in new, manually written code is strongly discouraged (see Section 15.1.2).

The `label` statement defines a unique label in a logical statement block (for example, function or control part) and the `goto` statement allows the execution to jump directly to the position of that label in the same statement block (Table 3.11).

To prevent the abuse of this mechanism, there exist a few restrictions on the use of the `goto` statement. Jumping out of or into functions, test cases, or the control part is not allowed. It is similarly forbidden to jump into a loop or into an `if-else` statement.

3.2.5 The `log` Statement

The `log` statement provides the means to write logging information to the logging interface of a test system. The format of the logged values is dependent on the logging interface implementation used with the test system. Examples of using the `log` statement with `charstring` literals and variable values are shown in Table 3.11. The complete set of array elements can be logged by specifying only the identifier of the array as shown in our example with `c_testNames`. Values of variables or array elements that have not been assigned a value prior to their logging will be logged as 'UNINITIALISED'. In addition, `log` statements can also contain constants and function parameters. Function instances can only be referenced if they have been defined with a `return` clause. Users should be careful when referencing function instances. It is not advised to log instances that influence test behaviour or contain loop constructs.

Next to these basic constructs, more advanced constructs such as test component references, templates, timers and TTCN-3 operations related to them can also be logged. Rather than listing examples for these constructs and operations at this point, we will provide them when each construct is introduced in the following chapters of this book.

3.2.6 The Control Part

The control part of a module is the entry point for execution in a TTCN-3 program and thus takes 'control' over the execution path. The control part contains an arbitrary number of control statements and function calls that reflect the dynamic behaviour of the test system. Usually, the role of the control part is to control and sequence the execution of test cases. Therefore, any direct communication with the SUT, verdict setting and the creation of dynamic configurations is explicitly forbidden during the execution of the control part. These operations can only be performed from within the associated called test case. In Table 3.12, the example uses the control part to call all the specified behaviour

Table 3.12 The complete host name look-up example

```
module hostLookup {
  const integer c_hostNameCount  := 5;
  const integer c_testNameCount  := 8;
  const integer c_maxLabelLength := 63;
  const charstring c_hostNames[c_hostNameCount] := { "www.altavista.com",
                                                     "localhost",
                                                     "www.nokia.com",
                                                     "www.sony.com",
                                                     "www.yahoo.com"};

  const charstring c_IPaddress[c_hostNameCount] := { "66.94.229.254",
                                                     "127.0.0.1",
                                                     "147.243.3.73",
                                                     "160.33.26.10",
                                                     "216.109.118.68" };

  const charstring c_testNames[c_testNameCount] := { "www.altavista.com",
                                                     "localhost",
                                                     "www.nokia.com",
 "www.thisisareallyexceedinglylargehostnamewhichisnotallowedbytheDNSstandard.com",
                                                     "www.zappo.com",
                                                     "www..com",
                                                     "www.sony.com",
                                                     "www.yahoo.com"};

  type integer HostIndex ( -1, 0 .. c_hostNameCount - 1 );

  function f_findHost( in charstring p_host ) return HostIndex {
    // the v_loop variable is only known within the scope of this loop
    for ( var integer v_loop := 0; v_loop < c_hostNameCount;
                      v_loop := v_loop + 1 ){
```

Table 3.12   (*continued*)

```
      if ( c_hostNames[v_loop]   == p_host ){
          return v_loop;
      }
   }
   return -1;
}

function f_resolve( in charstring p_host, out charstring p_ip )
return boolean {
   var HostIndex v_index := f_findHost( p_host );
   if ( v_index > = 0 ) {
      p_ip := c_IPaddress[v_index];
      return true;
   }
   return false;
}
function f_checkHostName( in charstring p_name ) return boolean {
   // check for names without dot or more than 63 characters between dots
   var integer v_lastDot := -1, v_count := 0, v_length := lengthof( p_name );
   var boolean v_longName := false, v_pairOfDots := false;
   while ( v_count < v_length and not v_longName and not v_pairOfDots ) {
      if ( p_name[v_count]   == "." ) {
         if ( ( v_count - v_lastDot ) > c_maxLabelLength + 1 ) {
            v_longName := true;
         } else if ( v_count == v_lastDot + 1 ) {
            v_pairOfDots := true;
         }
         v_lastDot := v_count;
      }
      v_count := v_count + 1;
   }
   return ( ( not v_longName ) and ( not v_pairOfDots ) )
}

control {
   var integer v_count := 0;
   var charstring v_ip;

   do {
      if ( f_checkHostName( c_testNames[v_count] ) ) {
         if ( f_resolve( c_testNames[v_count], v_ip ) ) {
            log( "host name resolved:", c_testNames[v_count] );
         }
         else {
            log( "host name could not be resolved:", c_testNames[v_count] );
         }
      }
      else {
         log( "invalid host name:", c_testNames[v_count] );
      }
      v_count := v_count + 1;
   } while ( v_count < c_testNameCount );
}
}
```

without the use of any test case and therefore without any communication – this is not the typical use for a control part.

The full example of the `hostLookUp` program is given in Table 3.12. The program tries to resolve the eight test names by searching for a matching host name in the defined table of data `c_hostNames`. In the control part, a `do-while` loop is used to cycle through all possible test names. In the loop, each test name is first checked for validity in the function `f_checkHostName`. If a test name has two successive dots or more than 63 characters between the dots, the name is not valid. In that case, the name is not looked up, instead `"invalid host name"` is logged to the output device. If the test name is valid, the function `f_resolveHostname` is called to look up the IP address in the table. Depending on the boolean result, an answer is logged and the next test name is processed.

3.2.7 Preprocessing Macros

The TTCN-3 core language supports a number of pre-defined precompiler-like macros that are called *Preprocessing Macros*. These macros can be used in the definitions or the control part to replace the macro occurrence by the `charstring` or `integer` value of the name the macro refers to at compile-time. Currently there are four different types of definition IDs supported in the language you will find in the course of this section. Each macro starts and ends with a '_'-character and between those characters the name of the macro is stated. Note that the *Preprocessing Macros* will not be replaced inside literal charstring values, in templates or in TTCN-3 comments.

3.2.7.1 The Module-Name Macro _MODULE_

If the (pre-)compiler parses the module-name macro it is replaced by the name of the TTCN-3 module it was found in. The name is inserted into the source code as a `charstring` value. In a module named 'U3Tester' the statement:

```
log( _MODULE_ );
```

will evaluate to:

```
log( "U3Tester" );
```

3.2.7.2 The File-Name Macro _FILE_

The _FILE_ macro is replaced with a charstring containing the absolute file name of the source file it has been found in including the path. For example,

```
log( _FILE_ );
```

will evaluate to:

```
log("/root/LTE_tester/v6.8/Connection_Tests/10_07_19/Test7.ttcn");
```

3.2.7.3   The File-Name Macro _BFILE_

This macro does the same as the last one except that the compiler replaces the macro with the file name of the source file without its path, that is, the basic filename. The exact format of the filename is dependent on the compiler implementation.

```
log( _BFILE_ );
```

will evaluate to:

```
log( "Test7.ttcn" );
```

3.2.7.4   The Line Macro _LINE_

The (pre-)compiler will exchange this macro with the integer value of the current line number in the file. A file starts with line number **1**. Each new line, including commented lines increase the line number by **1**.

3.2.7.5   The Scope Macro _SCOPE_

For the scope macro `_SCOPE_` we need to distinguish between named scopes and unnamed scopes. An unnamed scope is simply a statement block between a pair of curly brackets: {...}. A named scope is any kind of function, a control part or a module definitions part. As the module definitions part doesn't have its own scope delimiters the embracing scope – the module – in this case is used as the named scope.

For any unnamed scope the `_SCOPE_` macro is replaced by the next higher named basic scope. So if the `_SCOPE_` macro is used inside an unnamed statement block somewhere inside a function, the function name would replace the macro later on. The terms `component`, `testcase` and `altstep` will only be introduced at a later point in this book, but Table 3.13 represents what TTCN-3 calls named, basic scope units:

Table 3.13   Named Basic Scope Units

Scope unit	`charstring` replacement
Module	Module name
Control part	**'Control'**
Function	Function name
Component	Component name
Test case	Test case name
Altstep	Altstep name
Template	Template name
User defined named type	Type name

3.3   Summary

TTCN-3 source code is written in modules, which are separated into a definitions part, where data types, functions and constants are defined, and a control part that describes the dynamic behaviour of the tests that should run.

We have introduced functions as scope units that can be called from expressions, other functions or the control part. They may compute results depending on input parameters and may return a value, that is the result of this computation.

We have also introduced a small number of basic types and explained how to create new subtypes and arrays. We have seen that types and subtypes can be used to declare variables or constants.

We have shown that expressions in TTCN-3 can be written with different operators that combine compatible values, and they can be used to directly assign the result to variables. Lastly, we have seen that in TTCN-3 we can direct the program flow with conditional or loop statements.

4

Single Component TTCN-3

TTCN-3 has been developed for testing. A central notion of TTCN-3 is the concept of test cases. Test cases define which events are sent to the system under test (SUT) and how to react to events received from that system. In this chapter, we consider only message-based communication with the SUT and test cases executed on a single test component, that is *non-concurrent* TTCN-3. The subject of concurrent test cases running on several test components will be covered in Chapter 5. Chapter 6 will cover the area of procedure-based communication in TTCN-3 test cases.

In this chapter, the concepts for message-based communication and non-concurrent test cases are introduced with an example. We use this example to introduce the structural concepts of ports, components and timers. In addition, we also introduce the concepts of test cases and their verdicts and introduce the alt statement, altsteps and functions.

In this section, we use a ticket vending machine for regional trains as our example. The vending machine accepts 10 cent, 20 cent, 50 cent, 1 Euro and 2 Euro coins. Similarly, it can return coins of these values. These coins are modelled as a corresponding subset of integer, where the 1 and 2 Euro coins are modelled as 100 and 200, respectively. The geographical area served by the regional train system is partitioned into a set of zones. Three different tickets are offered: an 'A' ticket that allows travel within a single zone, a 'B' ticket that allows travel from one zone to a neighbouring zone and a 'C' ticket that allows travel within all zones. The tickets are modelled as a subtype of charstring. The corresponding type definitions are shown in Table 4.1.

The possible interactions between a traveller and a ticket vending machine are illustrated in Figure 4.1. A traveller can request a ticket. The traveller can put cash into the vending machine and can also receive cash back. Clearly, the ticket vending machine can also actually issue tickets.

The typical behaviour is that a traveller wants to buy a ticket. To do this, a traveller first requests the kind of ticket he wants to buy. Then he inserts the necessary amount of money. If the traveller has inserted a sufficient amount of money, then the ticket vending machine issues the ticket and returns the correct amount of change, if appropriate. This example will be extended throughout this section to illustrate specific features of TTCN-3.

An Introduction to TTCN-3, Second Edition.
Colin Willcock, Thomas Deiß, Stephan Tobies, Stefan Keil, Federico Engler and Stephan Schulz.
© 2011 John Wiley & Sons, Ltd. Published 2011 by John Wiley & Sons, Ltd.

Table 4.1 Types for the ticket vending machine

```
type integer Coin ( 10, 20, 50, 100, 200 );
type charstring Ticket ( "A", "B", "C" );
```

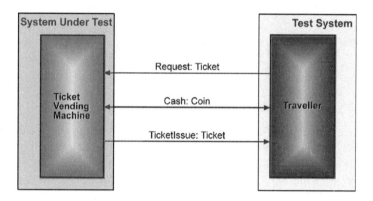

Figure 4.1 The ticket vending machine as the system under test.

The element we wish to test is the ticket vending machine. Therefore, the TTCN-3 test system needs to take the role of the traveller. Their interaction will be modelled as a series of TTCN-3 messages. The roles of the SUT and the test system are also shown in Figure 4.1.

4.1 Ports

Messages are sent via ports in TTCN-3. The messages sent via a port are delivered without delay to the corresponding recipient. Messages received at a port are stored in a message queue. Each port has its own message queue to hold the messages it receives. TTCN-3 does not put a limit on the length of the message queues. In an implementation of a TTCN-3 system, there can be practical limits.

A port type defines which messages can be sent via instances of this port type and, in the other direction, which messages can be received. In the ticket vending machine example, we define three port types as shown in Table 4.2.

The port type `Request` allows messages of type `Ticket` to be sent; no messages can be received via a port of this type. The directions `in` and `out` have to be seen from the test system point of view, in our case, a traveller. The direction `out` for the port type `Request` indicates that messages can be sent from the test system to the SUT when a traveller requests a ticket. Similarly, messages of type `Ticket` can be received via ports of the type `TicketIssue` when the ticket vending machine issues a ticket to a traveller.

Ports can be bidirectional, this can be seen from the port type `Cash` in the example, where messages of type `Coin` can be sent and received. A traveller has to pay for the ticket but can also receive change back. In this example, only messages of a single type can be sent and received over a single port, however, TTCN-3 also supports the

Table 4.2   Port types for the ticket vending machine

```
type port Request message {
  out Ticket
};

type port TicketIssue message {
  in Ticket
};

type port Cash message {
  inout Coin
};
```

Table 4.3   Ports with several message types

```
type port TVMPort message {
  inout Ticket;
  inout Coin
};
```

exchange of several message types over a single port. In Table 4.3, the port definitions from Table 4.2 have been combined to a single port definition TVMPort to illustrate how this can be done.

4.2 Components

The behaviour in a test system is executed on one or more test components. The ports of a given component describe its interface. Each component can also have its own local state, which consists of constants, variables and timers.

When defining a component type, the ports of the corresponding component instances are given by indicating their name and type. A test component can have several ports, and each port has its own queue to store received messages. In our example, three ports are used to describe the interface between the test system and the SUT. The first one is used for requesting a specific ticket from the ticket vending machine. The second port is used for issuing the ticket to the traveller, and the third port is used to exchange coins between the traveller and the ticket vending machine.

The resulting TTCN-3 code can be seen in Table 4.4. Remember that the port definitions indicate in which directions the messages can be sent.

Within a component type definition, it is also possible to have several ports of the same type, but the names of the ports must always be different.

Test components may have local constants, variables and timers, which are defined as part of the component type definition. Constants are defined with their type, name and value. Variables are defined with their type, name and optional initial value. A timer is defined by indicating its name together with an optional default duration. The unit for duration is seconds and must be given in the form of a float value.

Table 4.4 Interface of the ticket vending machine

```
type component TVMTester {
  port Request     pt_request;
  port TicketIssue pt_ticket;
  port Cash        pt_cash
};
```

Table 4.5 State of a test system

```
type component TSState {
  const integer c_maxCashAmount := 800;
  var   Ticket  v_ticketRequired;
  timer         t_inactive      := 5.0;

  port Request     pt_request;
  port TicketIssue pt_ticket;
  port Cash        pt_cash
};
```

In the example in Table 4.5, we extend the component type TVMState to also hold some state of the test system. The component type TSState defines a constant c_maxCashAmount with value 800 to describe the maximal amount of money a traveller is allowed to enter. The variable v_ticketRequired can store the kind of ticket the traveller requested; no initial value is given for this variable. The timer t_inactive can be used to test for inactivity from the traveller, which after some defined time should lead to the SUT aborting the current transaction. The default duration for this timer is 5.0 seconds. More information about timers will be given in Section 4.6.

Each instance of a component type will have its own instances of the ports, variables and timers. None of these entities are shared among several component instances.

4.3 Test Cases

In general, a test case is a behaviour description of how to stimulate the SUT and the expected reactions of the SUT to this stimulation. Depending on the reactions, a *verdict* can be assigned. For example, a test case can *pass* or *fail*. There are also other verdicts, which can be used to flag up a situation that made it impossible to decide if the test case passed or failed.

4.3.1 Main Test Component

On a more technical level, a test case in TTCN-3 defines the behaviour of the main test component. Because in this section we consider only test configurations with a single test component, the main test component is the only test component in such a test configuration. The interface of the test system towards the SUT – the test system interface

(TSI) – is, in such single test component configurations, completely defined by the ports of the main test component. Therefore, there is no need to define the TSI separately.

Each test case definition must refer to the component type on which the described behaviour is to be executed. This is done with the `runs on` clause in the test case definition. The test case `tc_empty` in Table 4.6 can be executed on an instance of the component type `TVMTester`. It has no parameters, as indicated by the empty pair of parentheses. As no statements are specified within its body, the test case will return immediately when invoked.

4.3.2 Test Case Verdict

Each test component has an implicitly defined variable of type `verdicttype`. This implicit variable is called the *local verdict* of a test component. The local verdict can be set by the operation `setverdict` and the value of the local verdict can be retrieved with the operation `getverdict`. The result of the `getverdict` operation can also be logged with the `log` statement. The initial value of the local verdict is `none`. In the test case `tc_empty` in Table 4.6, the verdict is not set, so that the verdict of this test case after its execution will remain `none`.

The test case `tc_pass` in Table 4.7 just sets the local verdict to `pass`. The test case `tc_fail` first sets the local verdict to `fail`, and thereafter sets it to `pass`. In this case, the value of the local verdict remains `fail` because the value of a local verdict cannot be improved in TTCN-3. Once the local verdict has been set to `fail`, it will remain `fail`. To state this in terms of testing: once an error in the SUT has been detected, it will be considered as erroneous.

The verdicts can be seen as ordered from `none` to `pass`, corresponding to the following relation: `none > pass > inconc > fail > error`. Note that in TTCN-3 setting the verdict `error` explicitly with the `setverdict` operation is not allowed. This verdict

Table 4.6 A test case with empty behaviour

```
testcase tc_empty () runs on TVMTester { };
```

Table 4.7 Test cases setting the local verdict

```
testcase tc_pass () runs on TVMTester {
  setverdict( pass );
};

testcase tc_fail () runs on TVMTester {
  var verdicttype v := getverdict;   // v == none
  setverdict( fail );
  v := getverdict;                   // v == fail
  setverdict( pass );
  v := getverdict;                   // v == fail
  log("The initial MTC verdict is: ", getverdict ); // logging via operation
};
```

Table 4.8 Test cases setting the local verdict and verdict reason

```
testcase tc_pass1 () runs on TVMTester {
  setverdict( pass, "verdict set to PASS" );
};

testcase tc_fail1 () runs on TVMTester {
  var verdicttype v := getverdict;    // v == none
  setverdict( fail, "The initial MTC verdict set to FAIL  v= ", v);
  v := getverdict;                      // v == fail
  setverdict( pass, "The initial MTC verdict set to PASS  v= ", v );
  v := getverdict;
  log("v= ", v );     // v == fail
};
```

is assigned by the run-time system whenever a run-time error, for example division by zero, occurs.

In addition to the local verdict each test component also has an implicit charstring variable which optionally allows the reason for the verdict assignment to be described. This variable is assigned via the setverdict operation. The reason string is passed as one or more optional parameters at the end of the setverdict operation parameter list. The rules and restrictions for these parameters are the same as those for the log statement. Every time the setverdict operation is called on a test component the reason string will be over-written. If no reason parameters are specified in the call the implicit reason string will be overwritten with the empty string. The reason string cannot be directly read or used within the TTCN-3 code, the information is available to the runtime execution system which will then handle it in a system specific manner. An example of the setverdict operation with reason information is shown in Table 4.8.

In addition to the local verdicts of the test components, there is the overall verdict of the test case. The overall verdict of a test case is the value it will implicitly return after its execution. Because we are considering only single test components here, the overall verdict coincides in our case with the local verdict of the main test component.

4.3.3 Test Case Invocation

In TTCN-3, test cases are invoked explicitly from the control part. A test case is invoked with the execute operation. The return value of the execute operation is the overall test case verdict. In the control part in Table 4.9, a variable of type verdicttype is defined. This variable is used to store the overall verdict of the test cases after they have terminated.

At first, the test case tc_empty defined in Table 4.6 is executed. As this does not have any behaviour defined, its local verdict is the initial value none. In our case, this is also the overall verdict; as a consequence, the variable v is assigned the value none. Note that the usual assignment operation – not the setverdict operation – is used to assign the value. The setverdict operation can only be used to assign values to the local verdict of a test component. Because the variable v is a normal user-defined variable, its value can be changed freely. The overwriting rule as described before applies only to

Table 4.9 Control part invoking test cases

```
control {
  var verdicttype v;
  v := execute( tc_empty () );      // v == none
  v := execute( tc_fail () );       // v == fail
  v := execute( tc_pass () );       // v == pass

  execute( tc_empty () );
};
```

the local verdict. Therefore, the value of v after executing the test case tc_pass has changed from fail to pass.

It is not necessary to store the result of the execute operation in a variable. It is also possible to just execute a test case and discard the return value. This can be seen from the second invocation of the test case tc_empty in Table 4.9.

The test system can be guarded against test cases that do not terminate by setting an upper bound on the execution time of test cases. This upper bound is given as an optional second parameter to the execute statement. If the test case does not terminate within the indicated time, the run-time system will force the test case to terminate and return error as overall verdict; see the example in Table 4.10. Even when some test case has the overall verdict error, further test cases as defined in the control part can still be executed. This is even possible if the verdict error has been set because of a run-time system error when executing the test case.

In Section 4.6, we will also show how timing restrictions can be handled within a test case. These two approaches, setting an upper bound in the control part or doing so in a test case itself, have their advantages and disadvantages. On the one hand, an upper bound in the control part has the advantage of being independent of potential errors in the test case. On the other hand, an upper bound in the test case allows specific behaviour to be defined, that is executed after the time bound has passed. It also allows the boundaries to be kept tight because individual safety margins can be kept smaller than an overall one.

In the previous examples, the control part looks mostly like a list of execute statements. Remember that all the statements as shown in Section 3.2 are still available and

Table 4.10 Upper bound on test case execution time

```
testcase tc_forever () runs on TVMTester {
  while ( true ) { /* loop forever */ }
  setverdict( pass )
}

control {
  var verdicttype v;
  v := execute( tc_forever(), 10.0 );   // seconds
                                        // v == error
}
```

these can be used, for example to describe repeated execution of test cases or to execute a test case dependent on the verdict of previous test cases.

4.3.4 Test Case Parameters

So far, we have only considered test cases without parameters. But similar to functions, test cases can also have parameters. Again, in, out and inout parameters are distinguished. The values for in parameters are passed by value. The values for out and inout parameters are passed by reference. This means that in the latter case, changes to the parameters in the test case are reflected in the control part. Note that if the verdict of a test case is error the values of out parameters are undefined.

The test case tc_parameter in Table 4.11 has a parameter of each of the three kinds. In the test case, the value of the in parameter p_1 is copied to the out parameter p_2 and the inout parameter p_3 is increased by one. Therefore, in the first execute statement the previously uninitialised variable v_2 is set to 1 and v_3 is set to 2. Both subsequent execute statements would cause run-time errors. The second execute statement has an expression for the inout parameter. In this case, it is not possible to pass the changed value back. Usually, this situation can be caught already in the semantic analysis of the TTCN-3 code and could be rejected as incorrect TTCN-3. The third execute statement has an uninitialised value for the inout parameter; this is also not allowed.

Parameters of test cases are one means to exchange data between subsequently invoked test cases. Static test configurations as described in Section 5.4 provide another means to exchange state information among sequentially executed test cases.

Table 4.11 A test case with parameters

```
testcase tc_parameter ( in    integer p_1,
                        out   integer p_2,
                        inout integer p_3 )
runs on TVMTester {
  p_2 := p_1;
  p_3 := p_3 + 1;
  setverdict( pass );
};

control {
  var integer v_1 := 1;
  var integer v_2;
  var integer v_3 := 1;
  var integer v_4;

  execute( tc_parameter( v_1, v_2, v_3 ) );  // v_2 == 1, v_3 == 2
  execute( tc_parameter( 1,   v_2, 3   ) );  // error, cannot assign value
                                             // to a constant expression
  execute( tc_parameter( v_1, v_2, v_4 ) );  // error, cannot pass an
                                             // uninitialized value as inout
                                             // parameter
};
```

4.3.5 Test Case Behaviour

A test component can have its own local constants, variables and timers. These entities are
brought into scope in a test case by the `runs` `on` clause in the test case definition. The
constants, variables and timers can then be used as local variables of the test case itself.
The component `TSState` shown in Table 4.5 has a variable `v_ticketRequired`. In
the test case `tc_purchaseA` in Table 4.12, this variable is set to `"A"` to indicate that
in this test case an `"A"` ticket should be purchased.

Note that in contrast to the local variable `v_ticketAmount` there is no declaration of
`v_ticketRequired` in the test case. This declaration has been given in the component
type definition. Independent of whether a variable is declared locally in the test case or
as a component variable, it can be used in the same way.

We will see later on that functions can also have a `runs` `on` clause and that the
constants, variables and timers of the associated component type then become accessible
within such functions. This means that from the viewpoint of test cases and functions
with a `runs` `on` clause, these constants, variables and timers are global entities. As with
any kind of global variables, these should be used cautiously.

Finally, see Table 4.12, TTCN-3 offers the predefined function `testcasename` which
allows the test case identifier to be accessed and to be used to influence test case behaviour.
The name is returned as a `charstring` value.

4.3.6 Test Case Termination

The execution of a test case terminates when its last statement has been executed. It does
not matter whether this is a `setverdict` operation or not. Even more, the execution of
a `setverdict` operation does not terminate the execution of a test case.

The `stop` operation allows a test case to be terminated at any point in its execution.
As a consequence of stopping the test case, the overall verdict is automatically returned
to the control part. In some cases the execution of a test case may come to a state, that is
considered as erroneous and from which it does not make sense to continue the computation. The operation `testcase.stop` sets the verdict to error and terminates test case
execution in a single operation. The optional arguments of the operation can be used to
indicate the reason of the termination similar to their usage in the `setverdict` operation.

Table 4.12 Using a component variable

```
testcase tc_purchaseA () runs on TSState {
  var integer v_ticketAmount;

  log ("Now executing test case: ", testcasename());

  v_ticketRequired := "A";
  v_ticketAmount   := 3;

  // more statements
};
```

4.4 Templates

Before we can finally start talking to the SUT, we present how to define the messages exchanged. In the case of sending a message to the SUT, the messages are just described as values of the corresponding type. For receiving messages, the situation might be more complicated if we wish to use matching expressions and not just straight values. In the example of the ticket vending machine, each coin inserted into the vending machine by the customer is a specific one. But the vending machine can choose how it will give the change and the traveller will accept any coins.

A *template* defines one or more values of a specific type. In this section, we will introduce only a restricted set of templates. They will be presented completely in Chapters 10 and 11.

The simplest way to define a template is as a single value, more precisely as a set consisting of a single value. Given the type Coin as a subset of integer, we define templates for each of the possible denominations, the templates a_coin10 to a_coin200 in Table 4.13.

Note that whenever a template can be used in TTCN-3, it is also possible to write a value explicitly. From this point of view, templates that define single values are not strictly necessary, but they are a convenient means to ensure consistency if the same value is used in several places. Note that templates and values are two different concepts. This is even true in the case of single-valued templates. A template can never be used directly in an expression.

Another simple way to define a template is a template containing all the values of a type. Such a template is defined by using the wildcard character '?' standing for any value. The template a_coinAny in Table 4.13 is defined in this way.

Between these two extremes of a single-valued template and all values of a given type, there is also the possibility to define a template consisting of several specific values, simply by enumerating them in the template definition as a *value list*. Because the values are explicitly enumerated, these templates define finite sets of values.

Table 4.13 Template definitions

```
// value as template
template Ticket a_ticketA := "A";

template Coin a_coin10   :=  10;
template Coin a_coin20   :=  20;
template Coin a_coin50   :=  50;
template Coin a_coin100  := 100;
template Coin a_coin200  := 200;

// any value template
template Coin a_coinAny := ?;

// value list template
template Coin a_smallCoins := ( 10, 20, 50 );
template Coin a_allCoins   := ( 10, 20, 50, 100, 200 );
```

Table 4.14 Templates as parameters

```
testcase tc_purchase ( in template Ticket p_ticket ) runs on TVMTester {
  // the test case body
}
```

The template `a_smallCoins` in Table 4.13 contains the three values 10, 20 and 50. The template `a_allCoins` contains all the possible values of the type `Coins`. Therefore, `a_allCoins` is equivalent to the template `a_coinAny`. This template was defined to contain all values of its defining type, which was again the type `Coins`.

Templates can be passed as `in` parameters to functions, test cases and so on. In this case, the parameter must be defined with the additional keyword `template`; otherwise, only values can be passed. The test case `tc_purchase` in Table 4.14 has the ticket to purchase as its parameter.

4.5 Message-Based Communication

To test the SUT, we need to exchange messages with it. The two most important operations to do this are the `send` operation and the `receive` operation. The `send` operation can be used to send a message to the SUT. The `receive` operation compares a received message against a template. If the message matches the template, then the message is removed from the message queue. As we have seen earlier, the messages are exchanged with the SUT via ports.

4.5.1 Send

The `send` operation transmits a message to the SUT via the specified port. The message is given by a template, which has to define a unique value.

In the ticket vending machine example, the traveller can request a specific ticket and then pay for it. Assume that an 'A' ticket costs 1.50 €. Then the request and the payment can be defined as in the test case `tc_purchaseA` in Table 4.15. The port `pt_request` is used to send the request for the ticket. The port `pt_cash` is used to pay for the ticket.

Whether the SUT can actually receive the message does not influence the execution of the `send` statement. As soon as the message can be delivered, the `send` statement is executed successfully and the execution proceeds. This means in this example that the verdict is set to `pass` independent of whether the SUT actually received the messages.

In the example above, all the messages are defined using templates of specific types and the ports are defined using the same types. For example, the port `pt_cash` allows messages of type `Coin` to be sent, which is the type of the template `a_coin50`. In addition to sending templates, it is also possible to specify an explicit value as a parameter to the `send` operation. In this case, it is not always possible to infer uniquely the type of the message, that is given as a value. In this case, the value can be preceded by a type name. In `send( integer:50 )`, the value 50 is of type `integer`, whereas in `send ( Coin:50 )`, it is of type `Coin`. It is important to know the correct type because values of different types might be encoded differently when they are sent to the SUT.

Table 4.15 Send statements

```
testcase tc_purchaseA () runs on TVMTester {
  // request the ticket
  pt_request.send( a_ticketA );

  // pay the ticket
  pt_cash.send( a_coin50 );
  pt_cash.send( a_coin100 );

  // to be continued

  setverdict( pass );
};
```

4.5.2 Receive

In addition to sending messages to the vending machine, we also want to see how the vending machine reacts to such requests. Therefore, we want to receive its reactions and check whether these are as expected. Messages in general are received by using the receive operation. This operation differs from the send operation both syntactically and semantically: Firstly, it can have a template as its parameter that describes more than a simple unique value. Secondly, the receive operation is a blocking operation. The receive operation compares the message at the head of the message queue of the indicated port with its parameter. This message is said to *match* the template if it is in the set of values described by the template. In cases where the template is a single value, then the message matches if it is exactly the same value. In cases where the template describes any value of a given type, then the message matches if it is of the correct type. And in cases where the template is defined as a value list template, then the message matches if it is one of the values in the template definitions.

Assume that the vending machine is returning cash after a traveller has requested a ticket and that at the head of the message queue is the value 100 of type Coin. Then the value matches the templates a_coin100, a_anyCoin and a_allCoins, which are defined in Table 4.13. But it does not match the templates a_coin200 and a_smallCoins.

Continuing the test case tc_purchaseA, let us assume that the SUT issues the correct ticket, that is an 'A' ticket. In our example, the message queue of port pt_ticket then contains the message 'A'. This message matches the template a_ticketA in the receive statement and the message is then removed from the message queue.

The receive operation is a blocking operation. This means that if the message queue does not contain a message, the execution of the test case is blocked at the receive statement. In the example, this means that the execution of the test case is blocked until the SUT eventually delivers an 'A' ticket. Note, however, that the receive statement is also blocking if the head of the message queue does not match the template in the receive statement. As the example is written at the moment, if the SUT erroneously delivers a 'B' ticket, then the execution of the test case is blocked forever. Even if the SUT would later deliver an 'A' ticket, the execution would not proceed because 'B' still

Table 4.16 Receive statement

```
testcase tc_purchaseA () runs on TVMTester {
  // request the ticket
  pt_request.send( a_ticketA );

  // pay the ticket
  pt_cash.send( a_coin50 );
  pt_cash.send( a_coin100 );

  pt_ticket.receive( a_ticketA );

  setverdict( pass );
};
```

sits at the top of the queue. We will see in Section 4.7 how this blocking can be handled in a more selective manner.

In the example in Table 4.16, it is clear which value has been received because the template describes a single value. But if the template describes more than a single value, then it is not clear which message has actually been received. To store the received message, it can be redirected to a variable: if pt is a port, a a template and v a variable, then value redirection is done with a statement pt.receive(a) -> value v. The received message is then stored in the variable v if the message matched. In the following chapter (Section 5.2.5) we show how parts of received messages can also be accessed.

In Table 4.17, we extend our example by actually overpaying for the ticket and checking that the traveller gets the correct amount of money back. The returned amount of change is stored in the variable v_returnedCash coin-by-coin and accumulated in the variable v_returnedCashAmount.

Note that in this example the variable v_returnedCash is of type Coin, whereas v_returnedCashAmount is of type integer. The variable v_returnedCash has to be of the same type as is used in the receive statement. The variable v_returnedCashAmount can hold other values, the sum of several coins does not have to coincide with the value of a single coin. Therefore, it has to be defined as an integer.

If one does not care at all about the received message, just that there is a message in the message queue, then the receive operation can be used without a parameter as in pt_cash.receive. Note that in this case value redirection is not possible.

In a similar way to the send operation, if the template is given explicitly as a value for the receive operation and the type of that value cannot be determined uniquely, then the type of the value can be given explicitly.

4.5.3 Check

If the first message in a port queue matches the template of a receive operation, then this message is removed from the queue. But sometimes it is convenient to leave the message in the port queue. This can be achieved by using the check operation; this

Table 4.17 Value redirection

```
testcase tc_purchaseAOverpay () runs on TVMTester {
  var Coin     v_returnedCash;
  var integer v_returnedCashAmount := 0;

  pt_request.send( a_ticketA );
  pt_cash.send( a_coin200 );

  pt_ticket.receive( a_ticketA );

  while ( v_returnedCashAmount < 50 ) {
    // wait for more cash
    pt_cash.receive( a_coinAny ) -> value v_returnedCash;
    v_returnedCashAmount := v_returnedCashAmount + v_returnedCash;
  };

  if ( v_returnedCashAmount == 50 ) {
    setverdict( pass );
  } else {
    // too much cash returned
    setverdict( fail );
  };
};
```

operation allows the inspection of the head of the message queue associated with a port without removing it. The check operation will block if there is no message in the queue or when the message at the head of the message queue does not match.

A check operation is written as an operation on a port with the receive statement and its parameters. If value redirection is used to store the message in a variable, this becomes a part of the parameter of the check statement. If the head of the message queue matches the template in the receive statement, the check statement is said to be successfully executed. If the message does not match, then the check statement will block.

Assume that at the head of the message queue of the port pt_Cash is the message with value 50. In the example in Table 4.18, the first check statement is executed successfully. Note that the message is not removed from the message queue. Hence, the second check statement is also executed successfully. The third check statement uses value redirection to store the message.

Table 4.18 Check statements

```
pt_cash.check( receive ) ;
pt_cash.check( receive( a_coin50 ) );

pt_cash.check( receive( a_coinAny ) -> value v_returnedCash );
```

Table 4.19   Receiving on several ports

```
any port.receive;
any port.receive( ? );

any port.receive( a_coinAny );

any port.receive( ? ) -> value v_returnedCash; // type error
```

4.5.4   Receive on Several Ports

Both the `receive` operation and the `check` operation can be applied to all ports of
a component at once. This is most useful to check for unexpected messages on several
ports in a single statement. Instead of the port name, the keywords `any port` are used
in the statements. It is not possible to use `any port` in a `send` statement because this
would lead to unpredictable behaviour of the test system.

Assume that all the statements in Table 4.19 are executed on an instance of the component type `TVMTester` that has two ports on which messages of different types can be
received. In the first `receive` statement, there is no template. This statement blocks until
there is a message in at least one of the message queues. If more than one message queue
contains a message, there is a random choice from which message queue the message
will be received. Note that it is not possible to determine from which port the message
was removed.

The second `receive` statement also blocks until there is a message in one of the
message queues. In the third statement, the `receive` operation blocks until there is a
message in the queue of port `pt_cash`; due to the type of the template `a_coinAny` this
can never match a message in the queue of port `pt_ticket`.

Value redirection can also be used together with `any port` but has to be used with
care. This can be seen in the fourth statement: the received message is redirected to the
variable `v_returnedCash`. Because this can lead to type incorrect assignments, this is
actually incorrect TTCN-3. To prevent such problems, explicit templates as in the third
statement should be used.

4.6   Timers

To describe timing properties, TTCN-3 provides timers. Timers can be started with arbitrary durations, and the execution can be blocked until a timer expires. In addition, there
are operations to stop a timer, check whether it is running and determine how much time
has passed since a timer was started.

One of the most common uses of timers in TTCN-3 is to guard against the inactivity
of the SUT. To achieve such a guard, the handling of the responses from the SUT must
be combined with the possible timing out of an inactivity timer. This will be described
in Section 4.7. In this section, we will introduce timers by expressing minimal separation
among subsequent events. As an example, consider the case when we want to express

that the user waits for 2 seconds after he has requested a ticket, before he starts paying by inserting coins into the ticket vending machine.

Durations for timers are given as non-negative `float` values in TTCN-3, and the unit of time is seconds, so a duration of 1 ms is given as 0.001. Note that it is the test system around TTCN-3 that actually decides how timers are implemented and what timer duration actually means. Most test system timer implementations provide a 'real' notion of time; however, it is also possible to implement a discrete notion of time, or whatever is considered appropriate in a specific testing context. For the sake of simplicity, we will consider that the timers implement a notion of wall clock time throughout this book.

Timers can be declared in the type definition of a component, in a test case, or in the control part of a module. Later on, we will also see that timers can be defined in functions and altsteps. Note: Timers declared and started in scope units such as functions cease to exist when the scope unit is left. In other words they do not contribute to the test behaviour once the scope unit is left. The declaration of a timer can provide a default duration. In this case, the timer can be started without giving a specific duration, with the default duration being automatically used. In the example in Table 4.20, the timer `t_ticket` is defined without a default duration. The timer `t_cash` is defined with an explicit default duration of 5.0 seconds.

A timer is started by the `start` operation with the timer duration as an optional parameter. If the timer has a default duration, the `start` operation does not need to have a duration as parameter. But if it has one, the value of the parameter overwrites the default duration.

In the example in Table 4.21, the timer `t_ticket` and the timer `t_cash` are used to specify that the traveller starts entering coins into the ticket vending machine no earlier than 2 seconds after requesting the ticket and that the traveller waits for 5 more seconds before entering the second coin. Here, the default duration of the timer `t_cash` was used. After starting the timers, the execution of the test case is blocked in the `timeout` operations until the timers expire.

The timer operations are written in a notation resembling object-oriented languages, like applying an operation to an object. But this is only a notational convention and by no means introduces object-oriented concepts into TTCN-3.

In cases where it is no longer necessary to have a timer running, a timer can also be stopped by using the `stop` operation. If a timer is stopped using the `stop` operation, this does not mean that it expired. A subsequent `timeout` operation will block waiting for a timeout that will never happen. Stopping an expired timer removes the corresponding timeout event; starting a running or expired timer is equivalent to first stopping and then restarting the timer.

There are two operations on timers that are less often used, one is to determine whether a timer is currently running and the second one is to determine how long a timer has been running. The `running` operation checks whether a timer is running, that is whether

Table 4.20 Timer declarations

```
timer t_ticket;
timer t_cash := 5.0;
```

Table 4.21   Timer start and expiration

```
testcase tc_purchaseA () runs on TVMTester {
  timer t_ticket;
  timer t_cash := 5.0;

  // request the ticket
  pt_request.send( a_ticketA );
  t_ticket.start( 2.0 );
  t_ticket.timeout;

  // pay the ticket
  pt_cash.send( a_coin50 );
  t_cash.start;
  t_cash.timeout;
  pt_cash.send( a_coin100 );

  pt_ticket.receive( a_ticketA );

  setverdict( pass );
};
```

it has been started, but not stopped or expired. The status of a timer is returned as a Boolean value, with `true` meaning that the timer is running. This operation can be used in expressions. Note that the `start`, `stop` and `timeout` operations do not return a value and hence cannot be used in an expression.

Similarly, the `read` operation returns the duration since a running timer was started. The duration is returned as a `float` value. For a non-running timer, the operation returns 0.0. This means that even if the timer was started and has expired, the `read` operation will return 0.0. The following example shows the use of the `read` and `running` operation for timers:

```
timer t;
var float v_elapsed := 0.0;
t.start( 2.0 );
if ( t.running ) {
  v_elapsed := t.read;
  log("Timer t is currently ", t, " with ", t.read,
      " seconds left!");
}
```

As shown in the previous example the state of a timer, for example inactive or running, and its elapsed time can also be logged to the logging interface.

Timers can be passed as parameters to altsteps or functions, as described in Sections 4.8 and 4.10. But they cannot be passed to test cases. Timers must be passed as `inout` parameters. It does not matter whether a timer is currently running or not when it is passed as a parameter.

4.7   Alt Statement

Both the timeout and receive operations presented in the previous sections are
blocking operations. If these operations are used as single stand-alone statements, as
soon as the execution of such an operation starts there is no possibility of proceeding
with the execution before a matching message has been received or the timer has expired.
Therefore, if at some point in the test execution we need to handle the situation where
several events might be received in any order, the stand-alone timeout and receive
operations are not suitable because this will require that the events are received exactly
in the specified order. Missing messages or messages that are received in unexpected
order can block the test case. For example, it is not possible to wait a certain period
for the reception of a certain message with stand-alone timeout and receive operations.
The stand-alone receive operation will always block indefinitely if no or an unexpected
message is received, regardless of any timer expiration.

The alt statement allows several blocking operations to be combined to overcome
the problems described above. The first of the blocking operations that can proceed
successfully determines how the execution continues.

The example in Table 4.22 extends the test case tc_purchaseA shown in Table 4.16
by restricting the time in which the ticket has to be delivered by the vending machine.
Thereby, it can be expressed that the test is only passed if the responses of the vending
machine occur in a timely manner. This also guards the test system against a broken
vending machine, which would not respond at all. Within the test case, after the ticket has

Table 4.22   The alt statement

```
testcase tc_purchaseTimely () runs on TVMTester {
  timer t_guard;

  // request the ticket
  pt_request.send( a_ticketA );

  // pay the ticket
  pt_cash.send( a_coin50 );
  pt_cash.send( a_coin100 );

  t_guard.start( 30.0 );

  alt {
    [] pt_ticket.receive( a_ticketA ) {
         t_guard.stop;
         setverdict( pass )
       };
    [] t_guard.timeout {
         setverdict( fail )
       }
  }
};
```

been paid for, the timer t_guard is started with a duration of 30.0 seconds. Thereafter, either a ticket is delivered by the vending machine within 30.0 seconds or not.

In cases where a ticket is delivered within 30.0 seconds, the first alternative in the alt statement is chosen, the timer is stopped and the verdict is set to pass. In cases where no ticket is delivered within 30.0 seconds, the timer expires. The first alternative cannot be chosen because no message has been received, but the second alternative is chosen because the timeout event has been received and the verdict is set to fail. There is no need to stop the timer in this case because it has already expired.

In both cases, the execution continues after the alt statement. In our case, there is no further statement after the alt statement, so the test case terminates.

The alternatives of an alt statement are evaluated top-down. Should, for some reason, the ticket be delivered after exactly 30 seconds, then both alternatives could potentially be chosen. In such cases, it is always the upper one, that is chosen. This top-down order can be used to give priority to special cases over general ones by describing them as the first alternatives.

In the example in Table 4.22, it is still not explicitly explained what happens if the ticket vending machine delivers an incorrect ticket. In such a case, the test case should clearly fail. To express this, we need to add an additional alternative to catch any delivered ticket. As this alternative follows the specific one for the correct ticket, this alternative will only be taken if a wrong ticket has been delivered; see Table 4.23 for an example.

It can happen that a message arrives or that a timer expires while some other alternatives are evaluated. To keep the top-down evaluation order and to avoid race conditions, the state of the test component is frozen before the alternatives are evaluated. Then, all alternatives

Table 4.23   An alt statement dealing with unexpected SUT behaviour

```
testcase tc_purchaseTimely () runs on TVMTester {
  timer t_guard;

  // as before

  t_guard.start( 30.0 );

  alt {
    [] pt_ticket.receive( a_ticketA ) {
         t_guard.stop;
         setverdict( pass )
       };
    [] pt_ticket.receive {
         t_guard.stop;
         setverdict( fail )
       };
    [] timeout.t_guard {
         setverdict( fail )
       }
  }
};
```

are evaluated against this frozen state or snapshot. Only if none of the alternatives can be chosen, then a new snapshot is taken and the evaluation starts again.

4.7.1 Boolean Guards

So far in our examples the square brackets in the `alt` statements have just indicated the beginning of a new alternative. However, more important than this syntactic functionality, these square brackets can also enclose a Boolean expression. Depending on the value of such a Boolean expression, an alternative will be evaluated or not. Only those alternatives whose guard evaluates to true are considered when looking for matching alternatives.

There is one special guard – `[else]` – for which no `timeout` or `receive` statement is needed. An `else` branch will always be chosen when none of the preceding alternatives has been selected.

In the example in Table 4.24, the first alternative will be considered only when the integer variable `x` is greater than 2. The second alternative will only be considered when the integer variable `x` is smaller than 0. Both of the alternatives will be chosen only when a message matching the corresponding template has been received. If neither the first nor second alternative is chosen, then the `else` branch is taken. For example, if `x` has the value -1 and the first message in the queue of port `pt_p` matches template `a_msg1` but not `a_msg2`, then neither the first nor the second alternative could be chosen. Instead, the `else` alternative would be chosen, resulting in the verdict `fail`.

Note that it is neither required that the Boolean guards are mutually exclusive nor that the offered alternatives are exhaustive. This means that several of the Boolean guards can evaluate to `true` in the same snapshot, but it is also allowed that none of them evaluates to `true`. In cases where several of the Boolean guards evaluate to true within a snapshot, the `receive` and timeout statements of these alternatives are considered top-down until the first successful evaluation determines which alternative will be chosen. In cases where none of the Boolean guards evaluates to `true`, there are two situations: if the guards are independent of the snapshot and do not change between subsequent evaluations due to side effects (like incrementing of component variables and so on) then this situation will not change by subsequent evaluations of the snapshot. This means that the `alt` statement would block forever, which is treated as a test case error. If the guards do depend on the current snapshot or can change by repeated evaluation, then a guard potentially may become true in the future, which means that the `alt` statement will have to be re-evaluated repeatedly. Although this might be seen as an interesting feature, we strongly advise such guard expressions be avoided. They may be treated differently by different TTCN-3 tools, and the resulting behaviour of the test system is

Table 4.24 Boolean guards

```
alt {
    [x > 2] pt_p.receive( a_msg1 ) { setverdict( pass ) };
    [x < 0] pt_p.receive( a_msg2 ) { setverdict( pass ) };
    [else]                         { setverdict( fail ) }
};
```

hard to understand. Indeed, the TTCN-3 standard prohibits the use of certain operations in guards that may change between subsequent evaluations of the same `alt` statement. Examples of such operations are the operations to check whether a timer or a component is currently running.

Functions that are called from the Boolean guard expressions in the `alt` statement must not alter the current snapshot. Again, TTCN-3 prohibits the use of certain operations in such functions. Prohibited operations are, for example the `receive` operation, changing a timer's status with the `start`, `stop` or `timeout` operations, or modifying component variables. The operations that cannot be used in the guards directly cannot be used in such functions either.

4.7.2 Repeat Statement

When an alternative has been selected, then the execution continues with the statements in the statement list of the chosen alternative. When all these statements have been executed, the execution continues after the `alt` statement. This means that the subsequent alternatives of the `alt` statement will not be considered, and the guard expressions will not be evaluated again. In cases where different situations are handled one after the other, this is just the behaviour we would wish. On the other hand, assume that we want to handle similar situations repeatedly, then we would like to 'restart' the `alt` statement instead of replicating its code or putting a loop with an auxiliary stop flag around it. The `repeat` statement of TTCN-3 allows exactly such descriptions to be written. When executing a `repeat` statement, the enclosing `alt` statement is evaluated again. A `repeat` statement actually can be used only within the alternatives of either an `alt` statement or within the alternatives of an altstep, which we will present in the subsequent section.

The example in Table 4.25 is a variation of the test case shown in Table 4.17. Both test cases check whether the ticket vending machine returns the correct amount of money if an 'A' ticket has been purchased and the customer paid with a 2 € coin. In the first test case in Table 4.17, a `for` loop was used to accept coins until enough cash has been returned. Here, we use an `alt` statement for the same purpose; after a coin has been returned by the ticket vending machine, it is checked whether enough cash or even too much cash has been returned. In cases where the money returned is still not enough, the evaluation of the `alt` statement is repeated. This means that the test case waits for another coin to be returned.

4.7.3 Alt Statements vs. Stand-Alone Blocking Statements

Alt statements allow several block operations to be grouped into a single statement, but of course it is also possible to have an `alt` statement with only a single alternative. In this case, the behaviour is exactly as if the single alternative would have been given without a surrounding `alt` statement. Indeed, the TTCN-3 standard specifies that stand-alone blocking operations will be treated by implicitly wrapping them into an `alt` statement. Table 4.26 gives an example for this expansion. We will come back to this implicit expansion of stand-alone `alt` statements in the discussion of default behaviour in Section 4.9.

Table 4.25   Repeat statement to await enough cash

```
testcase tc_purchaseAOverpayRepeat () runs on TVMTester {
  var Coin     v_returnedCash;
  var integer v_returnedCashAmount := 0;

  pt_request.send( a_ticketA );
  pt_cash.send( a_coin200 );

  pt_ticket.receive( a_ticketA );
  alt {
    [] pt_cash.receive( a_coinAny ) -> value v_returnedCash {
        v_returnedCashAmount := v_returnedCashAmount + v_returnedCash;

        if ( v_returnedCashAmount == 50 ) {
          setverdict( pass )       // correct amount of money returned
        }
        else if ( v_returnedCashAmount > 50 ) {
          setverdict( fail )       // too much money returned
        }
        else {
          repeat
        }                          // wait for more cash
    }
  }
};
```

Table 4.26   Expansion of stand-alone blocking statements

```
// standalone blocking operations
t_cash.timeout;
pt_ticket.receive ( a_ticketA );

// these are implicitly expanded to
alt {
  [] t_cash.timeout { }
}

alt {
  [] pt_ticket.receive ( a_ticketA ) { }
}
```

4.8   Altsteps

Several `alt` statements in a test suite might contain the same alternatives. In that case, the alternatives can be given a name and referred to in the `alt` statements to avoid code duplication. In TTCN-3, altsteps provide such a mechanism. Altsteps are similar to functions in that they can have parameters, but unlike functions, they also allow the description of guard expressions and `receive` and `timeout` operations.

Table 4.27 An altstep to wait for timer expiration

```
altstep alt_timeGuard ( inout timer p_t ) {
   [] p_t.timeout { setverdict( fail ) }
};
```

Table 4.28 Use of an altstep in an `alt` statement

```
t_guard.start( 30.0 );

alt {
  [] pt_ticket.receive( a_ticketA ) {
       t_guard.stop;
       setverdict( pass )
     };
  [] alt_timeGuard( t_guard );
}
```

The altstep in Table 4.27 has a timer as a parameter. It checks whether this timer has expired and if this is the case it sets the verdict to `fail`.

In the `alt` statement in Table 4.28, the previously defined altstep is used to describe that a ticket should be delivered within the specified time.

If the ticket has not been delivered, then in the altstep `alt_timeGuard` it is checked whether the timer `t_guard` has expired. If so, then the statements in this alternative of the altstep are executed. In our case, this means that the verdict is set to `fail`. There could also be further statements following the altstep in the `alt` statement, but in this example there are none.

If the altstep contains several alternatives, then these are evaluated one after the other as in an `alt` statement. If one of them can be chosen, the execution continues with the statements in the alternative's body. The execution continues until either the last statement in the selected alternative of the `altstep` is completed or an explicit **return** statement is executed. At this point the control returns to the invoking `alt` statement. If *any* of the alternatives in the altstep could be selected, then, after returning from the altstep, execution continues with the statements following the altstep invocation, which completes execution of the `alt` statement. Note: when using the **return** statement within an `altstep` definition no return value may be specified.

We have already seen that altsteps can have parameters. They also can have local definitions preceding the definition of the alternatives. Generally, such local variables are used to store received messages for processing them in the statements following a `receive` statement. In the `altstep` in Table 4.29, the variable `v_returnedCash` is such a local definition. It is initialised whenever the altstep is called and assigned a value when a coin is received. In this case, it is used to update the total amount of returned cash. An altstep can also have a `runs on` clause. In this case, the ports, timers, variables and constants of the component type can be used in the definition of the altstep in a similar way to test cases. In the altstep in Table 4.29, the port `pt_cash` can be used because it is a port of the component type `TVMTester`.

Table 4.29   Altstep with local definition

```
altstep alt_cash ( inout integer p_returnedCashAmount ) runs on TVMTester {
  var Coin v_returnedCash;

  [] pt_cash.receive( a_coinAny ) -> value v_returnedCash {
      p_returnedCashAmount := p_returnedCashAmount + v_returnedCash;

      if ( p_returnedCashAmount == 50 ) {
        setverdict( pass )              // correct amount of money returned
      }
      else if ( p_returnedCashAmount > 50 ) {
        setverdict( fail )              // too much money returned
      }
      else {
        repeat                          // wait for more cash
      }
  }
};
```

An altstep is always executed within a snapshot. Therefore, the initialisation of the local variables must not affect the snapshot. The restrictions on initialisation of the local variables of altsteps is the same as for the Boolean guards of the alt statements (see Section 4.7). In the case that local variables are used only to store received messages, the variables do not need to be initialised, therefore, there is no such problem.

Both altsteps that we have previously defined can be used to redefine the test case tc_purchaseAOverpay that was shown in Table 4.26. The changed version is shown in Table 4.30. The altstep alt_timeGuard is used to guard both the delivery of the ticket and the reception of the returned money. The reception of money is handled by calling the altstep alt_cash. This altstep contains a repeat statement. In this case, the inner of the two nested alt statements would be repeated.

The altstep when called can also optionally have a following statement block that will be executed after the functionality defined within the altstep if any of the alternatives within the altstep are triggered. An example is shown in Table 4.31. In this case if the guard timer times out triggering alt_timeGuard the statements after the altstep will be executed in this case resetting the verdict to inconclusive.

The return operation in an altstep can be used to end the execution of the altstep and return control to the enclosing alt statement. In this case the optional statement block following the altstep invocation would be executed. The break statement that was introduced for loop constructs can be used also for altsteps and for alt statements. When executed within an altstep the execution of the altstep will be terminated and the execution will continue after the enclosing alt statement, note that the optional statement block following the altstep invocation will not be executed. When the break statement is executed in the statement block of any alternative, then the subsequent statements in this block will be skipped and the execution will continue after the alt statement. In summary, the return statement leaves an altstep, the break statement leaves the enclosing alt statement.

Table 4.30   Use of several altsteps

```
testcase tc_purchaseAOverpayAltstep() runs on TVMTester {
  var integer v_returnedCashAmount := 0;
  timer t_guard;

  pt_request.send( a_ticketA );
  pt_cash.send( a_coin200 );
  t_guard.start( 30.0 );

  alt {
    [] pt_ticket.receive( a_ticketA ) {
        alt {
            // expect cash and return the amount of money
            [] alt_cash( v_returnedCashAmount )

            // guard against infinite waiting
            [] alt_timeGuard( t_guard )
        }
      };
    [] alt_timeGuard( t_guard )
  };
};
```

Table 4.31   Optional statement blocks in `altstep`-branches

```
t_guard.start( 30.0 );

alt {
  [] pt_ticket.receive( a_ticketA ) {
      t_guard.stop;
      setverdict( pass )
    };
  [] alt_timeGuard( t_guard ){ setverdict( inconc ) };
}
```

4.9   Default Altsteps

It is quite typical in test cases that many `alt` statements contain some alternatives to deal with unexpected SUT behaviour. A typical example is an alternative for receiving unexpected messages that have not been handled so far. Such an altstep is shown in Table 4.32.

To avoid adding such altsteps explicitly to each and every `alt` statement in the abstract test suite, it is possible to use altsteps as so-called *default* altsteps. An activated default altstep is added implicitly by a TTCN-3 run-time system at the end of each `alt` statement.

Altsteps that are used as default altsteps are defined just as other altsteps; in their definition, no distinction is made whether an altstep shall be used as a default or not. To use an altstep as a default, it simply has to be activated. This is done using the

Table 4.32   Altstep receiving any message

```
altstep alt_receiveAny() runs on TVMTester {
  [] any port.receive {
       setverdict( fail )
     };
};
```

activate operation. An activate statement has, as its parameter, an altstep together
with its actual parameters. Execution of an activate statement returns a reference to
the activated altstep; this reference can be used later on to deactivate an altstep, if wished.
The reference is of type default. The state of defaults can be written to the logging
interface by listing the applicable default variable as part of a log statement. The actual
format of state information in the log is however not standardised and tool dependent.

In the example in Table 4.33, the test case tc_purchaseAOverpay is extended
with a default to catch unexpected messages. Before the first alt statement, the altstep
alt_receiveAny, defined in Table 4.32, is activated and a reference is stored. This
altstep is now considered as a further alternative in each alt statement until the default
is deactivated. The deactivation is shown at the end of the test case.

The defaults are considered as alternatives after all the explicit alternatives from an
alt statement have been evaluated. This means that the alt statement from Table 4.33

Table 4.33   Default altsteps

```
testcase tc_purchaseAOverpayDefault() runs on TVMTester {
  var integer v_returnedCashAmount := 0;
  var default v_defaultRef;
  timer t_guard;

  pt_request.send( a_ticketA );
  pt_cash.send( a_coin200 );
  t_guard.start( 30.0 );

  v_defaultRef := activate( alt_receiveAny() );
  log( "INFO: Current state of activated defaults is: ", v_defaultref);

  alt {
    [] pt_ticket.receive( a_ticketA ) {
         alt {
           // expect cash and return the amount of money
           [] alt_cash( v_returnedCashAmount ) {};

           // guard against infinite waiting
           [] alt_timeGuard( t_guard ) {}
         }
       };
    [] alt_timeGuard( t_guard ) {}
  };
  deactivate( v_defaultRef );
};
```

Table 4.34 Alt statement with explicit defaults

```
alt {
   [] pt_ticket.receive( a_ticketA ) {
         alt {
            // expect cash and return the amount of money
            [] alt_cash( v_returnedCashAmount ) {};

            // guard against infinite waiting
            [] alt_timeGuard( t_guard ) {};

            // the default as explicit altstep
            [] alt_receiveAny() {}
         }
      };
   [] alt_timeGuard( t_guard ) {}

   // the default as explicit altstep
   [] alt_receiveAny() {}
};
```

corresponds to the one in Table 4.34, where the default altsteps have been added explicitly. Note that the default is added in both `alt` statements.

Similar to catching unexpected messages, it would be useful to use a default to guard against waiting infinitely for responses from the SUT. In the examples from Tables 4.33 and 4.34, we called the altstep `alt_timeGuard` in both `alt` statements. To replace these two explicit calls with a default, we have to cope with two problems. Firstly, if an altstep has call by reference parameters, then the referenced object might no longer exist when the altstep is executed as a default. Secondly, if several altsteps were activated as defaults, in which order will they be considered when looking for a matching alternative?

If an altstep has `out` or `inout` parameters, then it cannot be activated in TTCN-3 as a default. Such an altstep cannot be executed in a context where the variables to which the parameters refer are no longer available.

This rule is applicable to value and template parameters only. Timers and ports can be passed as `inout` parameters, and it is possible to use the altstep `as_timeGuard` as default.

In this example we use a different approach, instead of passing the timer as a parameter, we use a component timer. The timer `t_inactive` has been defined already in the component type `TSState` in Table 4.5. We define now the altstep `alt_inactive` to check whether this timer expired. This is shown in Table 4.35.

Table 4.35 Altstep to check for timeout of component timer

```
altstep alt_inactive( ) runs on TSState {
   [] t_inactive.timeout { setverdict( fail ) }
};
```

Note that we need the `runs` on clause to be able to refer to the component timer. Now we can add this default altstep to the test case and take care of the timeouts implicitly. Each `alt` statement can now be written with only one explicit alternative. Following the convention described in Section 4.7.3, such an `alt` statement with only a single alternative can be further shortened by dropping the surrounding `alt` statement, which is then implicitly present. This is also the case for `alt` statements that contain a single invocation of an altstep. An example is shown in Table 4.36 where the two definitions are equivalent.

This implicit expansion of stand-alone altsteps and blocking statements implies that default altsteps will also be considered when evaluating such stand-alone statements. The resulting test case is shown in Table 4.37. Note that we still need one explicit `alt` statement to describe that the altstep `alt_cash` will be called only if a_ticketA has been received, but not if one of the defaults was executed.

We can now have a look at the second question. In which order will the default altsteps be considered? One might think that the most obvious way is the order of the default activations. In our example, this would mean that the altstep `alt_receiveAny` would be the last default to be considered. But actually, the defaults are considered in reverse order

Table 4.36   Stand-alone altstep

```
alt {
   [] alt_cash( v_returnedCashAmount )
}

alt_cash( v_returnedCashAmount );
```

Table 4.37   Test case with multiple defaults

```
testcase tc_purchaseAOverpayDefaults() runs on TSState {
  var integer v_returnedCashAmount := 0;

  pt_request.send( a_ticketA );
  pt_cash.send( a_coin200 );
  t_inactive.start( 30.0 );

  activate( alt_inactive() );
  activate( alt_receiveAny() );

  alt {
    [] pt_ticket.receive( a_ticketA ) {
         // expect cash and return the amount of money
         alt_cash( v_returnedCashAmount );
       };
  };
};
```

of their activation. Using the reverse order of activation allows activating quite general altsteps at the beginning of a test case. Then in specific parts of the behaviour, more specific altsteps also can be activated and later be deactivated. Because these more specific altsteps are activated later, they will be considered before the general ones, which allow, for example reaction on some additional messages before `alt_receiveAny` causes test case failure for unexpected messages.

Although default altsteps can be used to get quite compact TTCN-3 codes, there is the danger that one loses track of the currently activated defaults, especially if one activates and deactivates altsteps frequently. This can lead to code, that is hard to understand, which in turn might be hard to maintain. Consider this as a word of warning and think carefully about how you actually write your abstract test suite.

4.10   Functions

In the previous section, we have seen how altsteps can be used to define several alternatives and how the altsteps can be called, either explicitly or implicitly as defaults. We have also seen in Section 3.1.9 how value-computing functions can be defined. Similar to such value-computing functions, it is also possible in TTCN-3 to define functions that express communication behaviour.

Such functions can contain any of the statements we have seen so far. They are more general than altsteps, which have to consist of a selection among alternatives on the topmost level. The functions that we are looking at here can, for example also start with a `send` statement.

The function f_payExact is an example of such a function. It is shown in Table 4.38. This function is intended to send messages of type `Coin` on the port `pt_cash` until the requested amount of money is actually 'sent'. For the sake of simplicity, we assume that the amount to pay is a multiple of 10 and only 10 cent coins are used for paying. The function has a single parameter and no return value. In the function body, we use the `return` statement to indicate the end of the function.

In the function body, the port `pt_cash` is used in a `send` statement. To be able to access this port, the function uses the `runs on` clause referring to the component type `TVMTester`. It is also possible to explicitly pass the port as an `inout` parameter to the function and to omit the `runs on` clause.

Table 4.38   Function to describe behaviour

```
function f_payExact ( in integer p_amount ) runs on TVMTester {
  for ( var integer v_alreadyPaid := 0;
        v_alreadyPaid < p_amount;
        v_alreadyPaid := v_alreadyPaid + 10 ) {
    pt_cash.send( a_coin10 )
  };

  return;
};
```

Table 4.39   Using a function defining behaviour

```
testcase tc_purchaseAFunction () runs on TVMTester {
  // request the ticket
  pt_request.send( a_ticketA );

  // pay the ticket
  f_payExact( 150 );

  // to be continued

  setverdict( pass );
};
```

Table 4.40   Recursive function

```
function f_pay ( in integer p_amount, inout Cash p_pt ) {
  var integer v_coin := 0;
  if        ( p_amount >= 200 ) { v_coin := 200 }
  else if ( p_amount >= 100 ) { v_coin := 100 }
  else if ( p_amount >=  50 ) { v_coin :=  50 }
  else if ( p_amount >=  20 ) { v_coin :=  20 }
  else if ( p_amount >=  10 ) { v_coin :=  10 }

  if ( v_coin > 0 ) {
    // pay largest possible coin
    p_pt.send( Coin:v_coin );

    // pay remaining amount
    f_pay( p_amount - v_coin, p_pt );
  }

  // completely paid
  return;
};
```

The function can be used in a test case as shown in Table 4.39, which is a variation of the test case tc_purchaseA that was shown in Table 4.15.

Although we have distinguished between value-computing functions and functions defining behaviour, this is not required by TTCN-3. It is possible to call a value-computing function from a behavioural function and vice versa. Even recursive calls are possible as shown in Table 4.40. There, the function is similar to the function f_payExact, but it does not use a for loop. Instead, it calls itself recursively until the requested amount of money has been paid. The port is passed as parameter; therefore, no runs on clause is necessary. In this example, we have also varied the method of payment, in so much as we have removed the limitation of only using 10 cents coins.

4.10.1 Restrictions on the Runs on Clause

After showing how functions can be used to define behaviour, we will now have a look at the restrictions that govern the components on which a function (or altstep) may be executed. We will also highlight the differences between functions and altsteps.

When using the `runs on` clause in a function definition, the function can execute on instances of the specified component type. With certain restrictions, it is also possible to execute the same function on instances of other, extended, component types. This allows, for example to call a function from test cases that execute on a different test component type. A function can be executed on instances of component types that are an *extension* of the one used in its `runs on` clause. The extended component type must have all the component timers, ports, constants and variables of the original type, with the exact same type and value, but it may have additional timers, ports, constants or variables. This means that the function `f_payExact` that has a `runs on` clause referring to the component type `TVMTester` can also be executed on component instances of type `TSState`. `TSState` has all the port definitions of `TVMTester` and some further variables and timers. Component extension is described in more detail in Section 5.4.

Similar to functions, an altstep with a `runs on` clause can also be used on other component types than the one specified in its `runs on` clause. Exactly the same restrictions as for functions apply.

There is a further, purely syntactical, restriction in TTCN-3 that applies to both functions and altsteps. If a function or altstep is defined *without* the `runs on` clause, then it is not allowed to call a function or altstep *with* a `runs on` clause. This ensures that illegal function and altstep invocations, which violate the restrictions from above, can already be caught statically, and not only at run time.

As we have seen, functions and altsteps have a number of similarities. Both can be used to define behaviour, they can have parameters, they can be defined with a `runs on` clause, and they can call other functions and altsteps. Even the call of a stand-alone altstep looks identical to a function call. In the following, we will summarise the differences between altsteps and functions.

- **Usage in alt statements:** Altsteps can be used in `alt` statements on the top level similar to `receive` and `timeout` statements. Functions can be used only in the statement lists of the alternatives or as part of the expressions in the Boolean guards.
- **Usage as default:** Altsteps without `out` or `inout` value or template parameters can be used as defaults. Functions cannot be used as defaults.
- **Top-level statements:** An altstep must consist of top level of alternatives similar to an `alt` statement. A function can start with any statement.
- **Initialisation of local variables:** The initialisation of the local variables in an altstep must be such that the current snapshot is not influenced. For functions, there are no restrictions on initialisations.
- **Return values:** A function can have a return value. An altstep cannot have a return value.

Therefore, an altstep is a more specialised way to describe behaviour than a function and it is mostly useful in the context of an `alt` statement.

4.11   Summary

In this section, we have presented how the behaviour of test cases can be described. We focused on test cases that are executed on a single test component, that is cases that are executed non-concurrently. We have explained that the ports of a test component define its interface. This means it defines which messages can be exchanged with the SUT.

We have briefly shown how templates are used to describe the messages that are sent to the SUT and the set of messages that can be received from the SUT. We introduced the corresponding `send` and `receive` statements. As the `receive` statement is blocking, we have shown how stand-alone `receive` statements can be extended to `alt` statements with several alternatives. In addition to `receive` statements, it is also possible to wait for the expiration of timers in an `alt` statement. Thereby, we can guard against a SUT that does not answer in a timely manner or does not answer at all.

An important concept when evaluating an `alt` statement is that all alternatives are evaluated against the same snapshot of the timer status and the message queues to avoid race conditions and inconsistencies.

5

Multi Component TTCN-3

Test systems often have to control several interfaces of the system under test (SUT). At each of these interfaces, the test system might have to take a different role towards the SUT. For example, in the extended Domain Name System (DNS) example in Section 2.1.9, the test system takes a number of roles, a DNS client, a domain name root server and a remote domain server. In TTCN-3, one or more ports can describe each interface towards the SUT, and each role can be reflected by a parallel test component.

In this chapter, we will present how test cases using several test components can be written. We will especially focus on how test components can be created and how the ports of the test components can be connected. The test components together with the connections among the ports are called *test configuration*. Test configurations in TTCN-3 are dynamic. This means that they can change while executing a test case. Compared to the use of a single test component as introduced in the previous chapter, there are several major differences. These differences are highlighted in the following.

- **Sequential versus concurrent behaviour:** The behaviour of each test component in isolation is sequential. In the single component case, the main test component (MTC) is the only component. Therefore, the behaviour of the whole test case is sequential. When several test components are used, their behaviour can be executed sequentially but also concurrently. The behaviour of all the test components contributes to the behaviour of the test case.
- **Combination of verdicts:** In the single component case, the verdict of the MTC becomes the overall verdict of the test case. In the multi component case, the local verdicts of all the test components contribute to the overall verdict of the test case.
- **Dynamic configurations:** In the single component case, there is one test component throughout a test case. In the multi component case, parallel test components can be created and terminated throughout the test case execution. Any test component can create and terminate other test components. In both the single and multi component

An Introduction to TTCN-3, Second Edition.
Colin Willcock, Thomas Deiß, Stephan Tobies, Stefan Keil, Federico Engler and Stephan Schulz.
© 2011 John Wiley & Sons, Ltd. Published 2011 by John Wiley & Sons, Ltd.

case, it is possible to change the mapping of component ports to the ports of the test system interface during execution.

- **Explicit test system interface:** In the single component case, it is optional to explicitly define the test system interface. If no test system interface is defined, the interface of the MTC is used as the definition of the test system interface. In the multi component case, it is mandatory to define a test system interface explicitly and to map the ports of test components to ports of the test system interface.
- **Data sharing:** On a single test component, data can be passed as parameters or the component variables can be used to pass data from one part of the code to another. Data shared by several test components has to be exchanged by explicitly passing it around in messages; there is no concept of global variables within a test case.
- **Test component execution:** TTCN-3 supports several different ways of test component creation. By default test components are restricted to execute only one single behaviour. The creation of alive components enables the execution of multiple behaviours sequentially which can help to simplify the specification of a multi component test case.

In this chapter, we start by presenting the extended DNS example in Section 5.1. In Section 5.2, we describe test components and their operations as the first part of test configurations. The second part of test configurations, the connections between ports and how these can be established, is presented in Section 5.3. Component type extensions, are introduced in Section 5.4. In Section 5.5, we shortly introduce some miscellaneous port operations. Finally, in Section 5.6 we show how TTCN-3 allows the use of addresses that refer to entities in the SUT.

5.1   Multi Component Test Case Example

In this section, we show a further test case for the DNS server that was introduced in Section 2.1.9. In this test case, we will check whether the SUT forwards the query in the correct way if it cannot resolve a host name itself. Three different entities are involved, as can be seen in the diagram in Figure 5.1, which is repeated from the introductory section.
   The three entities are as follows:

1. The DNS client that issues the original request to a DNS server to resolve a specific address.
2. The DNS root name server that knows which other server to ask if the query cannot be resolved locally.
3. The remote DNS server that probably is able to provide the answer.

If the local name server cannot resolve the host name 'www.nokia.com', then at first it will request the address of the name server responsible for the domain 'www.nokia.com'. The local name server will issue this request to a root name server. Once the local name server receives the reply from the root name server with the correct address 'ns.nokia.com', it will request the address for 'www.nokia.com' from the remote name server 'ns.nokia.com'. Lastly, when the local name server receives the reply from the

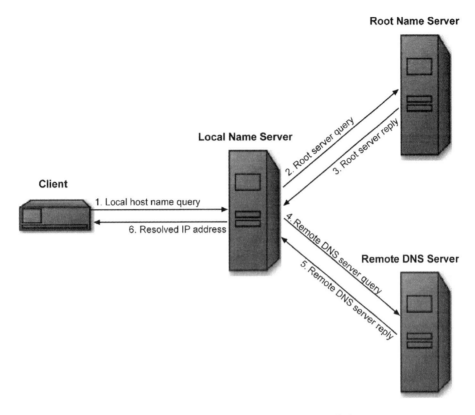

Figure 5.1   Entities in remote DNS name resolution.

remote name server, it will send the resolved address to the client. The sequence of message is shown in the message sequence chart (MSC) in Figure 5.2.

5.2   Test Components

In this section, we start by explaining the different types of test components. Here we consider the case that the lifetime of test components is limited to the duration of a test case. Thereafter, we explain the various operations related to test components.

5.2.1   Main Test Component and Test System Interface

In our example, we consider that the SUT is connected to the three network entities through three different interfaces. At each of these interfaces, DNS messages can be exchanged. Therefore, because the test system needs to take the role of all three network entities, we describe the interface of the test system towards the SUT by a component type that has three ports. The port type DNSPort has been already defined in Chapter 2 but is

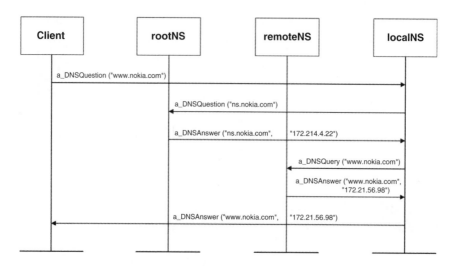

Figure 5.2   Message exchange for remote DNS name resolution.

Table 5.1   The test system interface for testing a DNS server

```
type port DNSPort message {
    inout DNSMessage
}

type component DNSServerTestSystemInterface {
    port DNSPort pt_client;
    port DNSPort pt_root;
    port DNSPort pt_remote;
}
```

repeated here for convenience. This interface component type is defined in Table 5.1. The ports are named after the role of the test component that should eventually send messages via them.

Note that the test system interface is not a test component on which statements are executed. It is just a set of ports. As such, it defines the interface of the test system towards the SUT and can be considered an abstract interface. It defines which messages or procedures can be exchanged with the environment of the test system in a test case, but it does not define how this is done on an implementation level.

In this test case, the test system will take the role of the three entities described above. Each of these entities will run on a different test component. The MTC will only coordinate the creation and execution of these three parallel test components and not interact with the SUT. Since coordination does not require communication with any other test component, the test component type `NoPortComponent` as defined in Table 5.2 will be sufficient for our MTC.

A multi component TTCN-3 test case starts with the execution of MTC behaviour as in the single component case. The type of this MTC is still specified by the `runs on`

Table 5.2 A type definition for a main test component

```
type component NoPortComponent { }
```

Table 5.3 A test case definition including a reference to a test system interface

```
testcase tc_remoteDNSResolution()
runs on NoPortComponent
system DNSServerTestSystemInterface {
    // ...
}
```

clause of the test case definition. The type of the test system interface is indicated by the `system` clause.

The definition of our testcase `tc_remoteDNSResolution` in Table 5.3 describes that MTC behaviour is executed on a test component of type `NoPortComponent` and uses a test system interface of type `DNSServerTestSystemInterface`.

5.2.2 Parallel Test Components

For each of the three entities, we will use a single parallel test component. Each of these entities has a single interface based on the DNS protocol towards the SUT. Correspondingly, each of the corresponding behaviours can be executed on a test component with a single port. In this case, we can reuse the same component type for the instantiation of all three entities. The corresponding definition is shown in Table 5.4; the component type `DNSEntity` has just a single port named `pt`.

5.2.3 Creation of Test Components

So far, we have talked about the definition of types for parallel test components, but we have not discussed how to actually create them. When executing TTCN-3, the MTC is created implicitly when a test case starts its execution. Parallel test components need to be created explicitly first by the MTC but they may in turn also create other parallel test components.

As shown in Table 5.5 parallel test components are created by using the `create` operation with a component type. Optionally, a logical name can also be associated with each test component instance which can be very useful for the analysis of execution logs of multi component test cases. During the creation of a test component any variable or

Table 5.4 The component type definition for parallel DNS test components

```
type component DNSEntity {
    port DNSPort pt;
}
```

Table 5.5 Establishment of a test configuration with multiple test components

```
testcase tc_remoteDNSResolution ()
runs on NoPortComponent
system DNSServerTestSystemInterface {
    var DNSEntity v_client;
    var DNSEntity v_root;
    var DNSEntity v_remote;

    v_client := DNSEntity.create("Client");
    v_root   := DNSEntity.create("Root");
    v_remote := DNSEntity.create("Remote");

    // ...
}
```

timer which is defined as part of the component type will be initialised and the ports of the component are enabled to engage in communication.

The create operation returns a reference to the created component instance which should be stored in a variable in order to be able to execute further operations on this instance such as the starting of behaviour which is discussed in the next section. Note that after the return of the respective create operations these three test components are just created, but no behaviour is executed on them. Also component references can be assigned to variables during their initialisation as shown in Table 5.6 and logged in log statements. Their representation in the test execution log is, however, tool dependent.

The test system configuration resulting from the creation of these test components is shown in Figure 5.3. The dotted arrows indicate that the parallel test components have been created by the MTC. Each of the boxes contains the assigned logical name to identify the role of the test component and the type of the component.

In Section 5.3, we will show how the ports of the parallel test components and those of the test system interface can be connected.

Table 5.6 DNS test component creation as alive components in variable initialisation

```
testcase tc_remoteDNSResolution ()
runs on NoPortComponent
system  DNSServerTestSystemInterface {
    var DNSEntity v_client := DNSEntity.create("Client") alive;
    var DNSEntity v_root   := DNSEntity.create("Root") alive;
    var DNSEntity v_remote := DNSEntity.create("Remote") alive;

    // ...
}
```

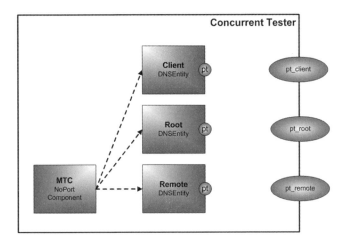

Figure 5.3   Test components in the DNS server test system configuration.

5.2.4   Alive Test Components

Parallel test components can be instantiated as components which can execute only one or multiple behaviours in their lifetime. This latter kind of component is called in TTCN-3 an alive component. The syntax for creating an alive component is shown in Table 5.6. Not only is the reuse of component ports when starting new behaviour on an alive component allowed, but also most of the component state is preserved. This means that component variables retain their values, and that messages in the port queues are preserved. The only state information not maintained are activated defaults. Since alive components can be used in the same manner as regular parallel test components, it is safe to always use the `alive` keyword in component creation. In Section 15.5 we illustrate in more detail how this concept can simplify the synchronisation of test components.

5.2.5   Component References

Component references refer to an instance of a component type, that is to the MTC or to a parallel test component. In the context of establishing test configurations some test components that are often referred to are the MTC, the test system interface and the test component, that is currently executing behaviour. But how can you get a reference to these component instances which are not explicitly created? For this purpose TTCN offers a few very useful pre-defined operations which are

- `mtc`: returns a reference to the MTC of a test case.
- `system`: returns a reference to the test system interface. This reference is needed when mapping ports of parallel test components to the test system interface.
- `self`: returns a reference to the test component on which this operation is executed.

The component reference `null` is a special value. It can be used to initialise variables for component references. However, using `null` as a component reference in a component operation will result in a run-time error.

Even after creating a test component, no statements are executed on it. The execution of statements on the test components only occurs after an explicit calling – or starting – of a function on the test components.

Before we present how these functions are started, we will have a look at the functions that are to be started. Table 5.7 shows two function definitions for the behaviour on the three parallel test components. The client behaviour is defined in the function `f_client`. This function is similar to the original test case, with the exception that the relevant information is passed on as parameters. A question message is sent and a specific answer is expected. The behaviour for the test components that act as root and remote DNS name server are both described by the function `f_server`. The same function can be used, but it will be called with different actual parameters. In this function, first, a question message is received and then the answer is sent back. The question message is stored and the identification, that is contained in the message is used in the answer message. This example shows how parts of messages can be accessed: Here the identification field

Table 5.7 Parallel DNS test component behaviours

```
function f_client( in Question        p_hostname,
                   in Answer           p_answer,
                   in Identification p_identification )
runs on DNSEntity {

  pt.send( a_DNSQuestion( p_identification, p_hostname ) );

  alt {
    [] pt.receive( a_DNSAnswer( p_identification, p_answer ) ) {
         setverdict( pass );
       }
  }
}

function f_server ( in Question p_hostname,
                    in Answer    p_address )
runs on DNSEntity {
  var Identification v_id;
  alt {
    [] pt.receive( a_DNSQuestion(?, p_hostname) )
         -> value v_id := DNSMessage.identification {
         pt.send( a_DNSAnswer( v_id, p_address ) );
         setverdict( pass );
       }
  }
}
```

received in the `query` message is stored in a local variable and used in the `reply` message. In case of a mismatch the variable continues to hold its original value, that is UNINITIALISED.

In the test case itself, after creating the test components and mapping the components, these functions are started by the MTC on the parallel test components and will then be executed in principle concurrently to each other.

In our example, we define and use constants to ensure consistency among the actual parameters and later on between the messages sent and expected. The corresponding part of the test case is shown in Table 5.8.

If the function called on the test component has formal parameters, then the actual parameter values must be given in the `start` operation. Values of any `out` or `inout` parameter as well as the return value of such a function are not accessible to the component invoking the `start` operation once the parallel test component terminates execution.

Table 5.8   Starting of parallel DNS test components

```
testcase tc_remoteDNSResolution ()
runs on NoPortComponent
system DNSServerTestSystemInterface {

  var DNSEntity v_client;
  var DNSEntity v_root;
  var DNSEntity v_remote;

  const Question       c_clientQuestion := "www.research.nokia.com";
  const Answer         c_clientAnswer   := "172.21.56.98";
  const Question       c_rootQuestion   := "ns.nokia.com";
  const Answer         c_rootAnswer     := "131.228.6.229";
  const Identification c_identification := 12345;

  timer t_guard;

  // create all parallel test components as alive components

  v_client := DNSEntity.create("Client") alive;
  v_root   := DNSEntity.create("Root") alive;
  v_remote := DNSEntity.create("Remote") alive;


  // ...

  // start the behaviour on the parallel test components
  v_client.start( f_client( c_clientQuestion, c_clientAnswer,
                            c_identification ) );
  v_root.start  ( f_server( c_rootQuestion,   c_rootAnswer ) );
  v_remote.start( f_server( c_clientQuestion, c_clientAnswer ) );

  // ...
}
```

Therefore, parameters cannot be used to pass information between test components – all communication of information has to be always performed using explicit communication such as message passing. Note that formal parameters of functions used in start operations may also not include any port or timer references.

Also, the function called must have a runs on clause. A function can be used in a start operation if the component type referred to in its runs on clause has at least the same component constants, variables, timers and ports as the component type of the referenced test component instance.

A further and important restriction is that it is only possible to start behaviour once on a test component which has been created without the alive keyword within a given test case. After executing a start statement for such a test component, any further start statement for this test component will result in a run-time error, even if the first function has already terminated.

5.2.6   Stopping Parallel Test Components

The behaviour on a test component terminates when the last statement of the function or test case executed on it has been executed. In addition to this obvious way of terminating the execution of component behaviour, a component can also be stopped explicitly using the stop operation. The stop operation can be called without a qualifying component reference. In this case, the test component instance on which this stop statement is executed terminates its behaviour.

The stop operation can also be qualified with a component reference. Using the component variables of the test case shown in Table 5.8, the statement client.stop would mean to stop the behaviour of the test component client. Also, it is possible to stop all parallel test components at once with the statement all component.stop. Note that this latter statement does not stop the MTC.

Whenever the MTC is stopped with mtc.stop, the test case as a whole terminates. As a consequence, any parallel test component which is still running is stopped implicitly. In this case the overall verdict is computed as usual. When a test case is stopped with the testcase.stop operation then the overall verdict is set to error. The test case stop operation has parameters similar to the log statement. These parameters can be used to indicate the reason for stopping.

Note that such an abrupt termination of a test case, as caused by mtc.stop or testcase.stop is often not desirable as it may leave the SUT or the test system in an unknown state and therefore prevent meaningful execution of any further test cases. As a rule it is better to design the test case so that the MTC should first wait for the other components to terminate on their own by using the done operation which is discussed in the following section, before calling mtc.stop.

In test cases with alive test components, the kill operation can be used to remove them from the test system. The difference in the effect of stop and kill operations in this case is that a stopped component can be restarted, whereas, a killed component can no longer be restarted. Another useful operation is the alive operation which returns true in case a parallel test component has not been killed yet and false otherwise, it is not a blocking operation.

5.2.7 Await Termination of Test Components

The `done` operation resembles the `timeout` operation for timers. It can be used to wait until a component has terminated; this means it is a blocking operation. Again, using the local variables of the test case shown in Table 5.8, the statement `v_client.done` would block until the component client has terminated its behaviour. The `done` operation can be used in a similar manner to `receive` statements in `alt` statements and altsteps. The `alive` operation returns `true` in case a parallel test component has not been killed yet, otherwise the operation returns `false` without blocking.

An example of such usage can be seen in the test case in Table 5.9. The statement `all component.done` used there actually means to wait until all parallel test components have terminated execution. In the example, there is also a timer used to guard against a component that does not terminate at all. Similarly, it is also possible to wait for any parallel test component to terminate by using the `any component.done` statement.

For alive test components the `killed` operation can be used in a similar way to the `done` operation to check if an alive test component has been removed from a test case via the invocation of a `kill` operation. Remember that killed test components cannot be started again. When the `killed` operation is used with non-alive test components it has the same effect as a `done` operation.

The execution status of the parallel test components is included in the snapshots that are used to evaluate the guards and receiving operations of the different alternatives. Therefore, all `done` statements in a single `alt` statement or altstep are evaluated against the same snapshot.

5.2.8 Checking Execution Status of Test Components

The `running` operation can be used to check whether a component is currently executing behaviour. This operation is again similar to the corresponding `running` operation on timers. This operation is non blocking and returns a Boolean value indicating the status of a test component. The value `true` is returned when the component has already started to execute behaviour and not yet terminated. The value `false` is returned if the component has been created already but not yet started to execute behaviour or if the component has already terminated. This operation can also be used as part of `log` statements to write its return value to the logging interface. An example of the `running` statement is shown in the code fragment in Table 5.10. In this example, `v_client` is a component reference. The code actually describes a form of actively waiting for the component `v_client` to terminate. It is checking once every second whether the component is still executing behaviour.

This looks quite similar to the statement `v_client.done`, with a time resolution of 1 second. However, in addition to this difference in timing, there is a further difference: If the component `v_client` has not yet started when the condition in the `while` statement is executed, the body of the `while` loop is not entered. The statement `v_client.done` would still block in this situation until the component actually started and terminated its behaviour.

Table 5.9 Waiting for DNS test component termination

```
testcase tc_remoteDNSResolution ()
runs on NoPortComponent
system DNSServerTestSystemInterface {

  // create all parallel test components as alive components
  // ...

  // start the behavior on the parallel test components
  // ...

  // wait until all parallel test components are done, at most 30 seconds
  t_guard.start( 30.0 );

  alt {
    var boolean v_alive;
    [] all component.done {
        t_guard.stop;
        // use verdicts of parallel test components
      };
    [] any component.killed {
        // if any PTC is killed then free all resources
        t_guard.stop;
        all component.kill;
        // remove all PTCs from test system which are still alive
        setverdict( inconc );
        v_alive := v_client.alive; // returns false
      };
    [] t_guard.timeout {
        // stop all PTCs (but new behavior could be started)
        all component.stop;
        setverdict( fail );
        v_alive := v_client.alive; // returns true
      }
    };
  // ...
}
```

Table 5.10 The running operation

```
while ( v_client.running ) {
  timer t;
  t.start( 1.0 );
  t.timeout;
}
```

The running operation can also be used to check whether any or all parallel test components are executing behaviour. The relevant statements are any component. running and all component.running, respectively.

5.2.9   Verdict Computation

Each of the parallel test components and the MTC have their own local verdict that can be accessed with the `setverdict` and `getverdict` operations (see Section 4.3.2). The overall verdict of a multi component test case is the 'worst' of the verdicts of the MTC and parallel test components. As an example, if one of the parallel test components has a verdict `fail` and all the other parallel test components and the MTC have the verdict `pass`, then the overall verdict of the test case will be `fail`.

Note that the MTC has its own local verdict. A test case can result in a `fail` verdict even when the verdict of the MTC is `pass`.

5.3   Mappings and Connections

In the previous section, we left open how the ports of the test components can be connected. TTCN-3 allows both to connect test components ports to other test component ports and to map a test component port to a port of the test system interface. Only after connecting or mapping a port is it possible to send or receive messages via this port. Note that two different terms are used in TTCN-3: *mappings* and *connections*.

5.3.1   Mappings

A port of a test component is *mapped* to a port of the test system interface by using the `map` operation. As an example, the `map` statements in Table 5.11 can be used to map the sole port of each of the parallel test components to the ports of the test system interface. In these statements, `v_client`, `v_root` and `v_remote` are component references to component instances. Each of these components has a port named `pt`. The test system interface, referred to by `system`, has three ports named `pt_client`, `pt_root` and `pt_remote`.

The resulting test system configuration is shown in Figure 5.4. The solid arrows indicate the established mappings. The dotted arrows indicate that the parallel test components have been created by the MTC.

Mapping the port `pt` of `v_client` to the port `pt_client` at the test system interface means that whenever on the component `v_client` a message is sent on the port `pt` this message is forwarded via the port `pt_client` towards the SUT. In the other direction, if the SUT sends a message, that is received by the test system on the port `pt_client` at the test system interface, then this message is forwarded to the component `v_client` and enqueued in the message queue of the port `pt`.

Table 5.11   Mapping of DNS component ports

```
map( v_client:pt,   system:pt_client );
map( v_root:pt,     system:pt_root );
map( v_remote:pt,   system:pt_remote );
```

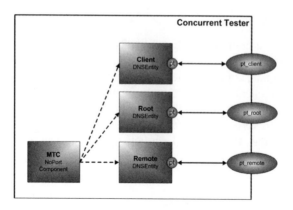

Figure 5.4 DNS test system configuration after mapping.

To enable the mappings to be changed at runtime, it must be possible to undo a mapping, which can be done with the operation unmap. The parameters are exactly the same as for the map operation. In the example shown in Table 5.12, the unmap statements are used at the end of the test case. In addition to unmapping ports one by one, it is possible to unmap all ports of a test component, all ports mapped to another port and even all mapped ports within the test case by a single operation. In this case the unmap operation will just have the port or test component as its single parameter. To unmap all ports of the test system the notation unmap(all component : all port) is used.

Although the mappings can be changed throughout test case execution, the mappings themselves must conform to certain rules. In our example, for any message that can be sent on the component v_client via the port pt, it must also be possible to send it via the associated test system interface port pt_client. This means that any message type, that is declared in the port type of the port pt as out must also be declared in the port type of pt_client as out.

In the opposite direction, for any message that can be received at the port pt_client it must also be possible to receive it at port pt. Both conditions imply that no message can get lost. At least, it cannot get lost without this situation being recognised: If this should occur at run time, then the TTCN-3 run-time system will signal a run-time error and set the local verdict of the relevant test component to error.

Table 5.12 Unmapping of DNS ports

```
  testcase tc_remoteDNSResolution ()
  runs on NoPortComponent
  system DNSServerTestSystemInterface {

    // ...
    unmap( v_client:pt,  system:pt_client );
    unmap( v_root:pt,    system:pt_root );
    unmap( v_remote:pt,  system:pt_remote );
}
```

5.3.2   Connections

The ports of test components can be connected directly to exchange messages between the two test components. The operation to connect two ports is `connect`. The parameters are similar to the `map` operation. However, whereas in the `map` operation one of the component references must be `system`, in a `connect` statement both references are referring to test components and not to the test system interface. The connection between ports can be changed throughout the execution of a test case; a connection is removed with the `disconnect` operation. As for unmapping several ports, all ports of a test component, all ports connected to one other port or all connected ports can be disconnected by a single `disconnect` operation.

Similar to the mapping of ports, there is a restriction on connections to ensure that each message sent can actually be delivered. If two ports are connected, for any message that can be sent via one of the ports, it must be possible to receive messages of this type at the other port and vice versa.

5.3.3   Many-to-One Mappings and Connections

So far, we have tacitly assumed that one port is mapped or connected to at most one port. In this case, it is quite easy to determine the receiving port when a message is sent. This is the port to which the sending port is mapped or connected. This means that for the sending component the recipient of a message is uniquely determined by the used port.

In the example in Section 5.1, each of the parallel test components had a specific role towards the SUT and each used a different port at the test system interface. Another scenario where parallel test components are useful is when several clients simultaneously request host name resolution from a DNS server. Extending this example further, each of the parallel test components measures how much time the server needs to answer its request. This time is sent from each of the parallel test components to the MTC. The ports needed to exchange these durations are called `pt_time` in our extended example. In this test configuration, each port `pt` of the parallel test components could be mapped to the same single test system interface port. Each of the ports `pt_time` of the parallel test components is connected to the port `pt_time` of the MTC. This test configuration is shown in Figure 5.5. The definitions of the ports are shown in Table 5.13.

TTCN-3 allows several ports to be mapped to one port. This is called a many-to-one mapping. Similarly, it is possible to connect several ports with one port. A test case definition could then begin as shown in Table 5.14.

A message of type `Time` can be sent from a parallel test component to the MTC simply by the statement `pt_time.send(duration)`, where `duration` could be a variable of type `Time`. A message of any of the parallel test components can be received on the MTC with the statement `pt_time.receive(?)`. It does not matter from which of the parallel test components the message was sent; the message would be received independent of its sender.

The actual sender of a message can be retrieved and stored in a similar way to the actual message received. Assume that `v_client` and `v_tempDuration` are two variables as defined in Table 5.15, then the `receive` statement can be used to receive any message of

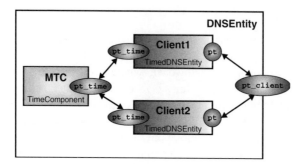

Figure 5.5   Test configuration with two client test components.

Table 5.13   Port and component definitions for transmitting time values

```
type float Time;

type port TimeIn  message { in  Time };
type port TimeOut message { out Time };

type component TimedComponent {
    port TimeIn pt_time
}

type component TimedDNSEntity {
    port TimeOut pt_time;
    port DNSPort pt
}
```

Table 5.14   Many-to-one connections and mappings

```
testcase tc_multipleClientDNSResolution ()
  runs on TimedComponent
  system DNSEntity {
    var TimedDNSEntity v_client1 := TimedDNSEntity.create("Client1");
    var TimedDNSEntity v_client2 := TimedDNSEntity.create("Client2");
    var TimedDNSEntity v_client3 := TimedDNSEntity.create("Client3");

    connect( v_client1:pt_time, mtc:pt_time );
    connect( v_client2:pt_time, mtc:pt_time );
    connect( v_client3:pt_time, mtc:pt_time );

    map( v_client1:pt, system:pt_client );
    map( v_client2:pt, system:pt_client );
    map( v_client3:pt, system:pt_client );

    // ...
}
```

Table 5.15   Storing value and sender of a message

```
var TimedDNSEntity v_client;
var Time v_tempDuration;

  // some code

  pt_time.receive( ? ) -> value v_tempDuration sender v_client;
```

type `Time` and store its actual value and the component reference of that test component from which the message was sent.

It is not only possible to store the sender of a message, but it is also possible to selectively receive messages from specific test components. To do so, a `receive` statement is extended by a `from` clause. In our example, let us assume that we want to receive a message of type `Time` from `v_client1` and store the value of the message and that we simply discard such a message from `v_client2` or `v_client3`. Such a selective `receive` statement is shown in Table 5.16.

To send a message to one of several components connected to the same port, a `send` statement can be extended with a `to` clause. This clause indicates the destination test component the message should be sent to. If the test component specified in the `to` clause does not exist or does not have a connection, this will cause a run-time system error. Considering the example shown in Table 5.15, a specific answer `a_answer` can be sent to the sending client by code as shown in Table 5.17. Note it is also possible to give a list of test components in the `from` or `to` clauses as shown in Tables 5.16 and 5.18. In the second case the message is sent to both test components, that is, this is a multicast `send` operation. To broadcast a message via all connected or mapped ports, the notation is `pt.send(...)` `to all component`.

Note that the use of component references in selectively receiving or sending messages imposes a further restriction on test configurations: A port can be connected or mapped to several other ports, but only if all these other ports belong to different test components. For example, a port cannot be connected to two ports on the same test component. In this situation, when sending a message it cannot be determined uniquely to which port the message will actually be delivered, because in the `send` statement only the destination component can be defined, but not the port at the destination component.

Table 5.16   Selective receive

```
alt {
  [] pt_time.receive( ? ) from v_client1 -> value v_tempDuration { };
  [] pt_time.receive      from (v_client2, v_client3) { repeat }
}
```

Table 5.17   Specifying the recipient of a message

```
pt_time.send( a_answer ) to v_client;
```

Table 5.18   Specifying a list of recipient's for a message

```
pt_time.send( a_answer ) to (v_client1, v_client3);
```

Table 5.19   Component definitions using extension

```
type component TimedDNSEntityI extends DNSEntity {
    port TimeOut pt_time; // "inherits" pt_dns from DNSEntity
}

type component TimeOutComponent {
    port TimeOut pt_time
}

type component TimedDNSEntityII extends DNSEntity, TimeOutComponent {
} // has same ports as TimedDNSEntityI
```

5.4   Component Type Extension

Once we feel comfortable with basic multi component test case specifications, we should consider how we can specify TTCN-3 definitions so that they can be reused efficiently. Whereas most aspects like naming conventions and modularisation are handled in Chapter 15 of this book, we introduce here the concept of extending component types which can be used to make component type definitions more compact and reusable.

Explicit component type extension is a fairly recent feature to the TTCN-3 language. The basic idea of extension is that component type definitions can be defined on the basis of others – in a way inheriting the constant, variable, timer and port definitions from the referenced component types. This allows a more compact test design approach where generic, highly reusable component 'base' types can be defined, for example for each specific test interface, from which then the types of components to be used in test cases are built. In conjunction, with component type compatibility (discussed in Chapter 9) these components can then also be used to define reusable, behavioural function definitions. One case where this construct could have been used is in our previous timed DNS resolution example as shown in Table 5.19.

Multiple component types can be referenced in component type extensions as well as types that extend another component type. Care has to be taken in the naming of constants, variables, timers and ports as part of the extension – they must not clash with definition name used in referenced type definitions. In addition, circular dependencies in component references are not allowed.

5.5   Miscellaneous Port Operations

Already when a test component is created, that is even before any behaviour has been started, all its ports can be used for enqueuing and transferring messages. In TTCN-3, the message transfer via ports can be stopped and restarted again. Messages via a stopped port will not be delivered and messages enqueued at a stopped port do not show up in the snapshots when evaluating alt statements. The operations to stop and restart message

Table 5.20   Logging of port status information

```
log( "INFO: The current status of port pt_time is ", pt_time );
```

transfer are `stop` and `start`. Additionally, it is possible to remove all enqueued messages at a port using the operation `clear`. The `log` operation can be used to log the status of a port, for example started or stopped, to the logging interface by listing the port instance identifier as shown in Table 5.20.

5.6   SUT Addresses

When many-to-one connections among ports are used, then several test components can exchange messages through a single port. Similarly, several entities in the SUT can exchange messages with the test system through a single port. References to such entities in the SUT are values of the type `address` and can be used in a similar way to component references in communication operations. This means that such addresses can be used in `to` and `from` clauses of the communication statements and can be stored in variables of type `address`. How the address type can be defined more specifically will be shown in Section 9.3.

5.7   Putting the Pieces Together

In this section we shall conclude this chapter by putting all the various pieces of the self-contained test case `tc_remoteDNSResolution` in their correct order. Remember that the `testcase` statement defined in Table 5.21 only describes the behaviour of the MTC. The behaviour of parallel test components is referenced here and defined in Table 5.7. After the local declarations in the test case, the parallel test components are created as alive components; thereafter, the ports of the test components are mapped or connected. Next, the behaviours on the parallel test components are started. After the third remote DNS test component is started, all three PTCs are active at the same time. Since in our example the handling of the different SUT interfaces are dependent on each other no explicit synchronisation of PTCs is needed. Note that even in the theoretical case that the SUT sends a message faster to the root test component than the test system starts up its behaviour, then this message is not lost but enqueued to its `pt` port and then processed once the root component starts. In more advanced test cases, it might be necessary to synchronise test component execution, for example when test components control completely independent SUT interfaces. We will have a closer look at synchronisation in Section 15.5.

After starting the behaviours on the parallel test components, the MTC blocks until all parallel test components have terminated or until 30 seconds have passed. At the end, the mappings of the ports are removed explicitly and the test case terminates.

It is also possible to use the concepts introduced in this chapter to write multi component tests with sequential component execution. One example is to start each test component individually and wait for its completion prior to starting the next one, that is to execute test components sequentially rather than concurrently as shown in Table 5.22.

Table 5.21 Complete definition of `tc_remoteDNSResolution` with concurrent PTC execution

```
testcase tc_remoteDNSResolution ()
runs on NoPortComponent
system DNSServerTestSystemInterface {
  var DNSEntity v_client;
  var DNSEntity v_root;
  var DNSEntity v_remote;

  const Question      c_clientQuestion := "www.research.nokia.com";
  const Answer        c_clientAnswer   := "172.21.56.98";
  const Question      c_rootQuestion   := "ns.nokia.com";
  const Answer        c_rootAnswer     := "131.228.6.229";
  const Identification c_identification := 12345;

  timer t_guard;

  // create all parallel test components (PTCs)
  v_client := DNSEntity.create("Client") alive;
  v_root   := DNSEntity.create("Root") alive;
  v_remote := DNSEntity.create("Remote") alive;

  // map the ports of all PTCs to the test system interface
  map( v_client:pt, system:pt_client );
  map( v_root:pt,   system:pt_root );
  map( v_remote:pt, system:pt_remote );

  // start the behaviour on all PTCs
  v_client.start( f_client( c_clientQuestion, c_clientAnswer,
                            c_identification ) );
  v_root.start ( f_server( c_rootQuestion,   c_rootAnswer ) );
  v_remote.start( f_server( c_clientQuestion,
                            c_clientAnswer ) );

  // wait until all PTCs are done, at most 30 seconds
  // after 30 seconds force stopping of all PTCs
  t_guard.start( 30.0 );

  alt {
    [] all component.done {
        t_guard.stop;
        // use verdicts of parallel test components
      };
    [] t_guard.timeout {
        all component.stop;
        setverdict( fail, "PTCs did not finish in 30 sec" );
        // only MTC verdict
      }
    };
```

Table 5.21   (*continued*)

```
    unmap( v_client:pt, system:pt_client );
    unmap( v_root:pt,   system:pt_root );
    unmap( v_remote:pt, system:pt_remote );
}
```

Table 5.22   Alternative test case specification using sequential test execution

```
testcase tc_remoteDNSResolutionSequential ()
runs on NoPortComponent
system DNSServerTestSystemInterface {
  var DNSEntity v_client, v_root, v_remote;

  const Question       c_clientQuestion := "www.research.nokia.com";
  const Answer         c_clientAnswer   := "172.21.56.98";
  const Question       c_rootQuestion   := "ns.nokia.com";
  const Answer         c_rootAnswer     := "131.228.6.229";
  const Identification c_identification := 12345;

  // create all parallel test components (PTCs)
  v_client := DNSEntity.create("Client") alive;
  v_root   := DNSEntity.create("Root") alive;
  v_remote := DNSEntity.create("Remote") alive;

  // map the ports of all PTCs to the test system interface
  map( v_client:pt, system:pt_client );
  map( v_root:pt,   system:pt_root );
  map( v_remote:pt, system:pt_remote );

  activate ( alt_timeoutHandler() );
  t_guard.start( 30.0 );

  // start the behaviour on PTCs one by one
  // due to active default every done implicitly is followed by a t_guard
  // timeout check
  v_client.start( f_clientSend( c_clientQuestion, c_identification ) );
  v_client.done;
  v_root.start(f_server( c_rootQuestion, c_rootAnswer ) );
  v_root.done;
  v_remote.start( f_server( c_clientQuestion, c_clientAnswer ) );
  v_remote.done;
  // this second start operation on the client requires an alive component
  v_client.start( f_clientReceive(c_clientAnswer,c_identification ) );
  v_client.done;

  t_guard.stop;
  deactivate();
  unmap( v_client:pt, system:pt_client );
  unmap( v_root:pt,   system:pt_root );
  unmap( v_remote:pt, system:pt_remote );
}
```

This style highlights more closely the message flow described in the original test purpose (see Figure 5.2) in its `testcase` statement. In our example we assume that `t_guard` timeout handling has been moved into an altstep definition which is activated as a default and also shows a first use scenario for alive components. In addition, we assume that in the case of a timeout a `stop` or `mtc.stop` operation is invoked. Note that `tc_remoteDNSResolutionSequential` not only realises the same test purpose as `tc_remoteDNSResolution` but it also reacts in the same manner to invalid SUT behaviour as the concurrent version of this test.

5.8 Summary

In this chapter we have introduced the specification of test cases with multiple test components. We considered the separate elements in isolation and then put the pieces together to define the self-contained test case `tc_remoteDNSResolution` in the last section. Although TTCN-3 does not require a specific order and style of declarations statements, as shown in Table 5.21, it has been proven useful in practice to group similar definitions together. We have shown how the `testcase` statement only describes the behaviour of the MTC. The behaviour of parallel test components is referenced from the function definitions. After the local declarations in the test case, the parallel test components are created; thereafter, the ports of the test components are mapped or connected. Next, the behaviours on the parallel test components are started. After starting the behaviours on the parallel test components, the MTC blocks until all parallel test components have terminated or until some guard timer has timed out. At the end, the mappings of the ports are removed explicitly and the test case terminates.

6

Procedure-Based Communication

So far, we have studied TTCN-3's mechanisms to test the system under test (SUT) via an asynchronous, message-based communication mechanism. This is the appropriate test approach in many applications and indeed has been the only communication paradigm that was available in TTCN-2 [25]. One of the major drivers behind the renewal of TTCN, which resulted in the standardisation of TTCN-3, was the extension of TTCN-based testing to new application areas. Many of these cannot be adequately modelled using message-based communication but rather require some form of procedure-based communication to be tested in a natural manner. Thus, a procedure-based communication paradigm was added to TTCN-3. In this chapter, we will study how systems that rely on procedure-based communication can be tested adequately with TTCN-3.

6.1 Procedure- versus Message-Based Communication

Message-based communication, which has been the dominating paradigm in the previous sections, is distinguished by the fact that communication is between equals: even in a client/server scenario, messages are sent and received using the same primitives (send, receive) regardless of the role of the communication partner. Only the semantics of the protocol create the distinction between clients and servers.

In contrast, procedure-based communication makes a clear distinction between these roles: for each communication act, there exists a distinct client that invokes (calls) a remote procedure, and a distinct server that processes this invocation and eventually returns a reply or, in erroneous conditions, raises an exception. All of this is done with specific communication primitives, depending on the current role of the communication partners.

There is a second important difference between message- and procedure-based communication. Message-based communication in TTCN-3 is asynchronous in nature, and the sender of a message will proceed with its behaviour before its message has been answered (or even has been received by its communication partner). In contrast, procedure-based communication is, in most cases, synchronous. The caller of a remote procedure will block until a reply has been returned or an exception has been raised. There are also

An Introduction to TTCN-3, Second Edition.
Colin Willcock, Thomas Deiß, Stephan Tobies, Stefan Keil, Federico Engler and Stephan Schulz.
© 2011 John Wiley & Sons, Ltd. Published 2011 by John Wiley & Sons, Ltd.

cases where the caller does not wait for the result of the invocation but rather attempts to collect the reply at a later point in time. Both forms of procedure-based communication are well supported by TTCN-3.

6.2   An Example – the Directory Service

Before we describe in detail how the procedure-based communication paradigm is supported by TTCN-3, we will introduce the example that we will use to demonstrate the relevant language constructs. For this, we will use a simplified directory server that associates keys (for example, names, login handles) with values (for example, addresses, passwords). When logging into the directory service, clients receive a reference to the directory, which they can use to look up keys. Certain privileged clients may also update the directory and insert new or update existing associations. A flag returned by the login procedure signals the client's status, that is if the client has read or read/write access to the directory. Exceptional error conditions cause exceptions, for example the lookup of non-existing keys or an attempt to update the directory without sufficient permissions.

Table 6.1 shows an Interface Definition Language (IDL) interface description for the directory service and its available methods. IDL [34] is a standardised language to specify interfaces for distributed, object-oriented systems and is used to specify CORBA-based systems. Much of TTCN-3's procedure-based communication mechanism has been shaped to allow for simple testing of CORBA-based systems, and hence an IDL-specified system is ideally suited to serve as an example in this chapter. At the same time, it is worth pointing out that TTCN-3's procedure-based communication mechanism is sufficiently rich to be used for the testing of arbitrary procedure-based communication without using CORBA or IDL.

First, let us have a closer look at how the directory service works: a client first uses the login method to acquire a reference from a well-known directory manager, which is responsible for checking the access rights to the directory. Once a handle to the directory has been acquired from the directory manager, this reference may be used to access the directory. The client's access capabilities are returned via the capabilities parameter. Read access is performed via the lookup method, which raises a NotFound exception if the queried key cannot be found. Write access is performed via the update method. If the value for an existing key is updated, the previous value is returned via the inout value parameter val of the update method, otherwise the empty string " " is returned via this parameter. An attempt to update the directory without write capabilities will raise a NotAllowed exception. A session is terminated by a call to logout, which invalidates the reference, that is further lookup or update invocation via this reference will raise the SessionExpired exception. For the sake of simplification we do not cover CORBA system exceptions in this example, see Section 9.5.2 to see how these are handled.

6.3   Procedure-Based Communication in TTCN-3

In the following, we will introduce TTCN-3's way of testing systems that use procedure-based communication, like the directory service outlined in Section 6.2. First, we need to define the interfaces that will be used to communicate with the SUT. This is analogous

Table 6.1   IDL description of the directory service

```
module DirectoryService {
  exception NotAllowed { string reason; };
  exception Rejected { string reason; };
  exception KeyNotFound { };
  exception SessionExpired { };

  enum Capabilities {
    e_reader, e_readerwriter
  };

  interface Directory {
    string lookup( in string key )
    raises ( KeyNotFound, SessionExpired );

    void update( in string key, inout string val )
    raises ( NotAllowed, SessionExpired );

    void logout();
  };

  interface DirectoryManager {
    Directory login( in string username,
                     in string password,
                     out Capabilities capabilities )
    raises ( Rejected );
  };
};
```

to the testing of message-based, asynchronous systems in which we started by defining the message and port types to be used.

We start with the explanation of signatures, which define the remote procedures to be used in testing. Table 6.2 lists the TTCN-3 signature definitions that capture the three methods of the `Directory` interface.

In general, a signature has a name, a possibly empty sequence of parameters with their types and passing modes, an optional return type, and a possibly empty list of exception

Table 6.2   The signature definitions for the directory interface

```
signature lookup( in charstring key ) return charstring
exception ( NotFound, SessionExpired );

signature update( in charstring key, inout charstring val )
exception ( NotAllowed, SessionExpired );

signature logout();
```

types. It is of course not accidental that a signature looks very similar to a function prototype – it represents a remotely invokable procedure; indeed, we will use these terms interchangeably. What has been added to normal functions is the possibility to specify exceptions, which are commonly used to indicate error conditions in many procedure-based distributed systems. We have already mentioned that the signatures specify the *types* of the exceptions, so to make the signatures from Table 6.2 fully defined, we need to define the types NotFound and SessionExpired. These exceptions carry no additional information other than their type, so for our purposes, trivial definitions can be used, for example empty records. It is possible to use arbitrary types for a signature's exception. In our IDL definition from Table 6.1, the exceptions Rejected and NotAllowed carry a string – the reason for the rejection – as the exception value. Again we use records, this time with a single charstring field, to model these exceptions in TTCN-3:

```
type record NotFound        { };
type record SessionExpired  { };
type record Rejected        { charstring reason };
type record NotAllowed      { charstring reason };
```

In a similar way to message-based communication, procedure-based communication takes place via ports. The port type definitions for our example are listed in Table 6.3. The fact that the Directory port type is used for procedure-based communication is indicated by the procedure keyword. Such port types may only specify signatures in their definition. The calling direction is indicated by the in, out and inout keywords. Signatures that are declared out or inout may be called via the port. In this case the test system plays the role of a client. For signatures that are declared in or inout, calls may be received. In this case the test system plays the role of a server. In our example we will let the test system act both in the client and the server role and have hence defined two port types, DirectoryClient for the case where a test component acts in the client role, DirectoryServer when a test component acts in the server role.

The restrictions placed on the possible connect and map operations for procedure-based ports are the same as those for message-based communication. Two procedure-based ports may be connected if the in or inout signature of one underlying port type is matched by a corresponding out or inout signature on the other underlying port type. The rule of thumb is that every signature that may be called from a client port must be callable on the server's port. A procedure-based port may be mapped to a procedure-based port on the test system interface (TSI), if each in or inout signature at the TSI port

Table 6.3 TTCN-3 port definitions for procedure-based communication

```
type port DirectoryClient procedure {
  out lookup, update, logout
}

type port DirectoryServer procedure {
  in lookup, update, logout
}
```

type is matched by an `in` or `inout` signature at the test component's port type, and if each `out` or `inout` signature at the test component's port type is matched by a corresponding `out` or `inout` signature at the TSI port type. As a rule of thumb: each outgoing call from the test component's port must be forwardable by the TSI port, and each receivable call at the TSI port must be forwardable to the test component.

For example, connecting a `DirectoryClient` port and a `DirectoryServer` based is legal, as well as mapping a `DirectoryClient` port at a test component to a `DirectoryClient` port at the TSI.

6.3.1   Non-Blocking Signatures

The underlying communication paradigm for procedure-based communication in TTCN-3 is that of synchronous communication via remote procedure calls (RPCs) – the caller of a signature blocks until the call returns. It is possible to deviate from this communication scheme, though, should it be necessary for testing purposes. One possibility is to declare non-blocking signatures; other possibilities will be described when we discuss the call invocation operations.

A signature may be declared as non-blocking if it does not specify a return type and has no `out` or `inout` parameters; `in` parameters are permitted. The invocation of such a remote procedure does not allow passing information back from the callee to the caller other than that the invocation has been received and possibly processed. One example for such a signature could be an unacknowledged version of the `logout` signature from the example above, which enables the client to log off from the directory service without waiting for an acknowledgement. A non-blocking signature is declared using the `noblock` keyword:

signature unackedLogout() **noblock;**

There are certain restrictions on non-blocking signatures when they are used in the communication operations, which we will discuss in Section 6.5.5.

6.4   Communication Operations

How does a typical invocation of the `lookup` signature look from the viewpoint of the two parties involved? The client sends a request to the directory, which is then received and processed by the server. Should the requested key be found in the directory, the server replies to the call and the client receives this reply. Should the key not be found, the server raises an exception, which needs to be caught by the client. For the moment, we will disregard how the client obtained a handle to the directory or how the directory is addressed; we will study these issues later.

From this description, we can identify three general modes of procedure-based communication:

- calling a signature (client to server);
- replying to a call (server to client); and
- indicating an exceptional condition (server to client).

We can also see that each of these modes involves a sending and a receiving party: the call is sent by the client and received by the server. The reply (or exception) is sent in the inverse direction from the server to the client.

6.5   Procedure-Based Communication on the Client Side

TTCN-3 allows tests from both the client's and the server's point of view to be specified. Consequently, there exist six communication operations in TTCN-3 that represent the possible combination of communication mode and sending or receiving side: `call` and `getcall`, `reply` and `getreply`, `raise` and `catch`. We will first discuss the operations that are used on the client side (`call`, `getreply` and `catch`) before covering those to be used on the server side.

6.5.1   The `call` Statement

The `call` statement is used to invoke a signature on a port declared for procedure-based communication. It specifies the signature of the procedure to call and the actual values for the signature parameters are given in the form of a template:

```
// pt is a mapped or connected port of type DirectoryClient
pt.call( lookup:{"password of John"} ) { ... }
```

The port specified in the statement must be connected or mapped. Its underlying type has to be of `procedure` kind, and must list the called signature among its `out` signatures. As for message-based communication, it is possible to define explicit templates to be used in procedure-based communication. These will be covered in Chapter 10. For our goal here – to introduce procedure-based communication – inline templates will suffice. Inline templates specify all procedure parameters directly after the signature identifier within the `call` statement. Specific values must be specified for each `in` or `inout` parameter of the signature. Any values of `out` parameters will be ignored in a `call` statement, a hyphen '–' can be used to avoid specifying a value for such a parameter.

From the minimal example in Table 6.4, you can see that the `call` operation does not handle the return value of the procedure invocation. Instead, any `call` statement for a signature that has not been declared non-blocking *must* have a body that handles the different possible results of the call. For example, to test if we can set John's password successfully to `"pa$$w0rd"` and if John indeed had previously no password set (indicated by an empty string being returned in `val`), one could use the call in Table 6.4.

Let us study this `call` operation and its body in some more detail. It is a typical example of a synchronous, blocking `call` operation without timeout. Asynchronous procedure invocation and timed procedure invocations will be treated in the following sections. In this simplest form of the `call` operation, the call is followed by a body – structured like an `alt` statement body – that enumerates different possible outcomes of the call. The body of the `call` statement is restricted to only deal with the possible outcomes of the call. It is not permitted to use altsteps or an else branch in the body. In addition, any operation that guards an alternative in the `call` statement

Table 6.4   Setting John's password

```
pt.call( update:{"password of John", "pa$$w0rd"} ) {
    [] pt.getreply( update:{-, ""} ) {
        // password successfully set, no previous password
        setverdict( pass );
    }
    [] pt.getreply( update:{-, ?} ) {
        // password successfully set, but old password existed
        setverdict( fail );
    }
    // exception handling should follow here
};
```

body must refer to the same port and the same signature that has been used in the `call` operation. We will now study the available guard operations that are available inside the `call` statement body in more detail.

Typically, procedure-based communication is unicast. Nevertheless, TTCN-3 allows multicast and broadcast procedure-based communication with a similar syntax as for message-based communication. In a multicast or broadcast `call` operation all the replies have to be handled in the body of the `call` statement. We will not explain multicast or broadcast procedure-based communication because it is rather untypical.

6.5.2 The `getreply` Operation

The first alternative in Table 6.4 specifies the SUT's desired behaviour. We expect to get a reply to our `update` invocation with the second parameter set to the empty string because no password has been set before this call:

```
[] pt.getreply( update:{-, ""} ) { ... }
```

Note that the `getreply` operation specifies the port on which we are waiting for a reply. This *must* be the same port as the initial call was issued on. In a similar way, the template used to specify the expect reply *must* have the same signature as the initial call.

In our example, the first parameter is unconstrained as it is an `in` parameter and any constraint would be ignored by the `getreply` operation. Knowing this, we have chosen to write '–', we could have equally written 'password of John' or even 'password of Janet' because an `in` parameter will not affect the outcome of the matching. For the second, `inout` parameter, we require that the empty string is set upon the return of the invocation. This empty string indicates that no previous value for John's password had been stored in the directory. Any other value is covered by the second alternative and leads to a `fail` verdict.

The `update` operation does not have a return value. Success is indicated by the fact that the procedure returns without raising an exception, and the relevant information is passed back via the second parameter. Of course, it is also possible to specify constraints on the return value, if the called procedure has a declared return type. For example, if

Table 6.5 Specifying constraints for the return value

```
pt.call( lookup:{"password of John"} ) {
   [] pt.getreply( lookup:{-} value charstring:"pa$$w0rd" ) {
      setverdict( pass );
   }

   [] pt.getreply( lookup:{-} value charstring:? ) {
      setverdict( fail );
   }

   // exception handling should follow here
}
```

we would like to check that, after a successful update, the set value is indeed returned by the lookup method, we could do this as shown in Table 6.5.

As can be seen from this example, constraints for the return value of a remote procedure must be specified as part of the getreply operation and are prefixed with the keyword value. In our example, the first alternative requires that the return value of the lookup method must be "pa$$w0rd", the second alternative deals with any other return value. It is not mandatory to specify constraints for a return value, even when the procedure has a return type, so the second alternative could be written more succinctly as:

```
[] pt.getreply(lookup:{-}) { ... }
```

6.5.2.1 Value Redirection for the getreply Operation

Like for the receive operation, the getreply and analogously the getcall and the catch operations do not perform value binding during the matching. Even when variables are used to specify constraints to be used in the matching of the received reply, these will not be bound to the actual values in the reply, but instead the variables' current values are used in the matching. Using undefined or partially defined variables will cause a test case error.

If specific values need to be known, for example to act differently depending on returned values, then it can be accessed by parameter redirection. For example, to access the actual value of the val parameter in the update procedure, one could use value redirection as shown in Table 6.6.

Similar to value or sender redirection for the receive operation, the parameter redirection is prefixed with an arrow '->' and followed by the keyword param. After the keyword, there is then a comma-separated list with one entry per parameter of the signature. Each out or inout parameter may be redirected to a variable of the corresponding type. For in variables, and for variables that do not need to be redirected, '-' can be used as a placeholder.

There exists a second notation for parameter redirection, that is similar to the field-assignment list notation for structure values. This alternate notation is shown in Table 6.7. Further redirections could be put into the same list, separated by commas. Note that the

Table 6.6 Parameter redirection

```
var charstring v_oldValue;

pt.call( ... ) {

   [] pt.getreply( update:{-,?} ) -> param ( -,v_oldValue ) {
        if ( v_oldValue == "" ) { }
      }
}
```

Table 6.7 Assignment notation for parameter redirection

```
var charstring v_oldvalue;
pt.call( ... ) {
   [] pt.getreply( update:{-,?} ) -> param (v_oldValue := val ) { }
}
```

assignments are of the structure 'variable := parameter' because it is the variable that receives the assignment of the actual parameter value. The big difference to the first notation is that there is no need to have one entry per parameter. In the case that a signature has many parameters and we only wish to redirect one or two, this second notation is far more concise. Regardless of which notation is used, each variable used for parameter redirection must have exactly the same type as the parameter, that is redirected to it.

For remote procedures with a return value, the actual returned value can also be redirected. This is done using exactly the same syntax as value redirection for the receive operation, which is shown in Table 6.8. The variable used in the return value redirection must exactly have the declared return type of the signature. For a signature that has both out or inout parameters and a return value, both forms of redirection can be combined. In this case, the value redirection must precede the param redirection.

6.5.3 The catch Operation

Let us now come back to the initial call example. If you are familiar with object-oriented programming or remote procedure invocation, you probably will have noticed

Table 6.8 Return value redirection

```
var charstring v_returnVal;

pt.call( ... ) {

   [] pt.getreply( lookup:{-} value ? ) ->  value v_returnVal { }

}
```

Table 6.9 Catching exceptions

```
pt.call( update:{"password of John", "pa$$w0rd"} ) {

  // getreply alternatives omitted here

  [] pt.catch( update, NotAllowed:? ) {
       setverdict( fail );
     }
  [] pt.catch( update, SessionExpired:? ) {
       setverdict( fail );
     }
}
```

that we have so far not considered how to handle exceptions. Considering the procedure
update, an exception may be generated when either we do not have the permission
to update the directory or we try to access the directory via a handle that has been
invalidated. In the former case, the NotAllowed exception will be raised, in the latter a
SessionExpired exception. We can deal with these exceptions in the body of the call
statement using the catch operation as shown in Table 6.9.

You can see that the catch operation specifies a port and the signature type, plus a
template that constrains the exception value that shall be caught. Unlike the getreply
statement, it is indeed only the signature identifier, that is given, not a signature
template. The reason for this is that a procedure that causes an exception will not
return normally, so it does not make sense to specify constraints for the out parameters,
inout parameters or the return value. Hence, the signature identifier is sufficient in this
case. When used in the body of a call statement, the catch operation must specify
the same port and the same signature type as the call operation.

When non-trivial types are used for the exceptions, then we can further constrain the
exceptions that we want to catch, for example if we want to test that an exception is
raised for weak passwords that fail to comply with the directories security standard. Such
a condition could be indicated by a NotAllowed exception that contains a corresponding
error message. The example code to catch such an exception is shown in Table 6.10.

Value redirection when catching exceptions is possible and uses the same syntax as
return value redirection for the getreply operation as shown in Table 6.11. Note this
redirects the value of the exception, not the return value of the associated function, which
is not accessible because the function did not return normally due to the exception.

Finally, sender redirection is available for all receiving procedure-based communication
operations (getreply, catch and getcall) using the same syntax as for message-
based communication. For more information see Section 5.3.3.

6.5.4 On Defaults, Deadlocks and Timed Invocations

So far, we have not explained how the evaluation of a call statement body is done
during test execution, and mostly this is indeed unnecessary because it works exactly
like a normal alt statement. A port has an associated queue into which all replies

Table 6.10   Constraining exception information

```
pt.call( update:{"password of John", "password"} ) {
  [] pt.catch( update, NotAllowed:{reason := "weak password"} ) {
       setverdict( pass );
     }
  [] pt.catch( update, NotAllowed:? ) {
       // e.g., "permission denied"
       setverdict( fail );
     }
  [] pt.catch( update, SessionExpired:? ) {
       setverdict( fail );
     }
  [] pt.getreply( update:? ) {
       setverdict( fail );
     }
}
```

Table 6.11   Value redirection with catch

```
var NotAllowed v_reason;
pt.call ( ... ) {
  [] pt.catch( update, NotAllowed:? ) -> value v_reason   { }
}
```

and exceptions for remote procedure invocations on that port are inserted as they arrive – either at the TSI, in the case of mapped ports, or from other components, in the case of connected ports.

There is one important difference though. During the evaluation of the body of a `call` statement, all active defaults are ignored. When none of the alternatives of the `call` statement body have matched, the execution of the current component blocks until something happens that makes a re-evaluation of the alternatives necessary.

This means there is a danger of deadlocking the test system or at least one test component. For example, the initial `call` statement from Table 6.4 without exception handling will deadlock should an exception be raised during the evaluation of the remote procedure. Even with the exception handling added, deadlocks may occur if the SUT loses the call and does not generate either a reply or an exception. It is possible to guard against this form of deadlock by setting a timeout period for the `call` operation. If no reply or exception is received within this period, the test system will generate a timeout exception. An example of this is shown in Table 6.12.

As usual, the timeout period is specified by a `float` value measured in seconds. As shown, the timeout exception is handled with the `catch` operation, using the keyword `timeout`. Use of this construct is only permitted if a timeout has been specified in the `call` operation. Note that the timeout is generated by the TTCN-3 system and is not related to the underlying RPC mechanism that is used by the test system adapter to communicate with the SUT. Therefore, it is guaranteed that the timeout exception will

Table 6.12   Timed calls and catching a timeout exception

```
pt.call( update:{"password of John", "pa$$w0rd"}, 5.0 ) {
   // getreply and exception should be handled first here

   [] pt.catch( timeout ) {
      setverdict( inconc );
   }
}
```

reliably terminate the `call` statement even in those cases in which the call yields a result, that is not covered by any other alternative.

It should be noted that special care needs be taken to ensure that late replies, which are received after a timeout exception has terminated a `call` statement, are cleared from the procedure port queue. If such replies are not cleared, they may be mistaken for the replies from subsequent calls or might even cause a deadlock, when they block the front of the queue and replies to subsequent calls can only be queued after them. See Section 6.5.5.1 for a more detailed discussion of this matter.

6.5.5   Non-Blocking Use of the `call` Operation

So far in this chapter we have considered the typical form of a remote procedure invocation and the way to deal with the possible outcomes of the call. Now we will go on to look at those cases in which the `call` operation is used to invoke remote procedures in a non-blocking manner.

We have already mentioned that remote procedures without `out` parameters, `inout` parameters or a return value can be declared to be non-blocking with the `noblock` keyword. Note that it is still possible to specify exceptions in non-blocking signatures. For example, the directory service could provide a function that allows for fast, unacknowledged updates of the directory:

```
signature bulk_update( in charstring key,
                       in charstring val ) noblock
exception ( NotAllowed, SessionExpired );
```

A `call` statement for such a non-blocking signature does not need to have a body and must not specify a timeout. The example in Table 6.13 shows how the `call` operation can be used for a non-blocking signature, for example to perform a number of subsequent updates without first checking if the updates have been successful.

It is also possible to invoke those signatures that have not been declared non-blocking in a non-blocking manner. In this case, instead of a timeout value for the call duration, the keyword `nowait` is specified. To perform a similar sequence of updates to the last example, but this time using the `update` method, the following loop could be used:

```
for ( var integer i := 0; i < 3; i := i + 1 ) {
   pt.call( update:{c_keys[i], c_values[i]}, nowait );
}
```

Table 6.13 Performing non-blocking calls

```
const charstring c_keys[3] := > {
  "password of John", "password of Janet", "password of Spikey"
};

const charstring c_values[3] := {
  "pa$$w0rd", "t1m3w4rp", "pocahontas"
};

// bulk update
for ( var integer i := 0; i < 3; i := i + 1 ) {
  pt.call( bulk_update:{c_keys[i], c_values[i]} );
}
```

In both cases, the `call` operation works asynchronously and returns immediately after the call has been initiated. Eventual replies or exceptions caused by these invocations are queued at the port and have to be processed at a later point in time.

6.5.5.1   Dealing with Results from Non-Blocking Calls

Of course, there comes a point where we want to check that our previous `bulk_update` invocations have not triggered any exceptions. This can be done with the `catch` operation that, in addition to the already shown `call` statement body, may also be used stand-alone or inside a normal `alt` statement. For example, to check that no exception is raised within 5 seconds as a result of the invocations of `bulk_update`, the code in Table 6.14 could be used.

Similarly, we may want to check that all invocations of the `lookup` method have successfully completed within 5 seconds. This can be done using `getreply` operations in an `alt` statement like the one in Table 6.15.

Table 6.14 Catching exceptions from non-blocking calls

```
timer t_guard;
t_guard.start( 5.0 );

alt {
  [] pt.catch( update, NotAllowed:? ) {
       setverdict( fail );
     }
  [] pt.catch( update, SessionExpired:? ) {
       setverdict( fail );
     }
  [] t_guard.timeout {
       setverdict( pass );
     }
}
t_guard.stop;
```

Table 6.15   Getting replies from non-blocking functions

```
timer t_guard;
var integer v_replyCount := 0;
t_guard.start( 5.0 );

alt {
  [v_replyCount < 2] pt.getreply( update:{-,?} ) {
      v_replyCount := v_replyCount + 1;
      repeat;
    }
  [v_replyCount == 2] pt.getreply( update:{-,?} ) {
      setverdict( pass );
    }

  [] pt.catch( update, NotAllowed:? ) {
      setverdict( fail );
    }

  [] pt.catch( update, SessionExpired:? ) {
      setverdict( fail );
    }
  [] t_guard.timeout {
      setverdict( inconc );
    }
}
t_guard.stop;
```

Of course, any port used for a stand-alone getreply or catch operation must be connected or mapped and have an underlying port type, that is of procedure kind and lists the used signatures among its out signatures. For all blocking call statements, this was guaranteed because otherwise the initial call statement, for which the same requirements exist, would have been illegal.

Special care should be taken that all pending replies and exceptions are eventually cleared from the port queue because otherwise they will interfere with subsequent calls. If, for example a NotAllowed exception for a non-blocking invocation of bulk_update is inserted into the queue but not removed by a matching catch operation, then it will block the head of the queue and a subsequent, synchronous call to update or lookup will either deadlock or time out (if a timeout value has been specified for the call). This is because any reply or exception for this subsequent call will be queued behind the exception, that is already in the queue and will hence not be inspected when examining the alternatives in the call statement body.

Any pending replies or exceptions can be explicitly cleared from a port queue using the clear operation. It is also possible to use the unconstrained version of the catch and getreply operations, which are analogous to the unconstrained form of the receive statement. This could be done as shown in Table 6.16.

Table 6.16   Getting rid of pending replies and exceptions

```
alt {
  [] pt.getreply {
      repeat;
    }
  [] pt.catch {
      repeat;
    }
  [else] {
      // empty queue - exit the alt statement
    }
}
```

Table 6.17   Setting fail verdict for any exception

```
altstep alt_failOnException() {
  [] any port.catch {
      setverdict( fail );
    }
}
```

The stand-alone versions of getreply and catch may also be used to react on all incoming events on all procedure-based ports. This is done using the any port keyword instead of a port name. For example, to set the fail verdict for any exception, that is raised on any port, the altstep in Table 6.17 could be activated as a default. When using this approach to deal with exceptions, remember that any default is ignored during the evaluation of a blocking call statement.

6.6   Procedure-Based Communication on the Server Side

So far, we have studied the communication operations that can be used on the client side of procedure-based communication. Often, we will also need a TTCN-3 test system to take over the server role. For example, this is necessary when testing a client implementation or when procedure-based communication is used for inter-component communication *inside* the TTCN-3 test system. The TTCN-3 operations for the server side are getcall to receive incoming procedure invocations and reply to dispatch the corresponding invocation result. Finally, raise can be used to send exceptions back to the invoking client. The concepts and syntax for the server-side communication are very similar to those for client-side communication that we have introduced in the previous sections. Therefore, we will keep our discussion short on the server side to avoid unnecessary repetition.

6.6.1   The getcall Operation

The getcall operation is used to accept incoming calls from other components or the SUT. For example, to accept incoming calls to the lookup or update method, the alt statement in Table 6.18 could be used.

Table 6.18 Accepting calls in TTCN-3

```
// pt is a mapped or connected port of type DirectoryServer (see Table 6.3)

alt {
  [] pt.getcall( lookup:{?} ) {
        // deal with the lookup procedure
     }
  [] pt.getcall( update:{?,?} ) {
        // deal with the update procedure
     }
}
```

Table 6.19 Constraining accepted calls

```
template charstring a_weakPassword
        := pattern "(password)|(root)|(admin)";

pt.getcall( update:{"master password", a_weakPassword} );
```

The operation specifies a port on which to listen, which must be connected or mapped and must have an underlying port type, that is of `procedure` kind and lists the expected signature among its `in` signatures. Furthermore, a template for the signature of the incoming call has to be specified, this can be either an inline or explicit template. For our purposes, inline templates are used. For more information on explicit templates for procedure-based communication see Chapters 10 and 11.

The signature template constrains both the type of the incoming call and the accepted parameters. In the example shown in Table 6.18, calls are accepted for the `lookup` and the `update` signatures, but not for the `logout` signature. There are no constraints specified on the parameter values because we want to accept arbitrary calls to these procedures. We could, of course, choose to specify additional constraints, but only for the `in` and `inout` parameters as these are only parameters passed from client to server. For example, to check that the 'master password' is not set to one of the weak passwords `"password"`, `"root"` or `"admin"`, the `getcall` operation in Table 6.19 could be used as an additional alternative.

Like `in` parameters for `getreply`, `out` parameters are ignored when matching an incoming call in the `getcall` operation. It is allowed to use the hyphen '–' for those parameters.

6.6.1.1 Value Redirection for the `getcall` Operation

It is possible to redirect the incoming `in` and `inout` parameters for a matching `getcall` operation to variables. This can be done by using the `param` keyword followed by a list of variables that shall contain the actual values of the incoming parameters.

The example in Table 6.20 shows both available syntactical forms of parameter redirection. The first redirection uses the explicit assignment to bind the variable `v_actualKey`

Table 6.20   Redirecting incoming parameters

```
var charstring v_actualKey;
var charstring v_actualValue;

alt {
  [] pt.getcall( lookup:{?} ) -> param ( v_actualKey := key ) {
       // try to find v_actualKey in the directory and
       // create a reply or exception
     }

  [] pt.getcall( update:{?,?} ) -> param ( v_actualKey, v_actualValue ) {
       // process the update and create an appropriate reply
     }
}
```

to the actual value of the parameter `key`. The second redirection lists all variables that shall be bound to the actual values of the parameters. For parameters that shall not be redirected, the '−' symbol can be used.

6.6.2 The `reply` Operation

So far, we have seen how incoming calls can be accepted by a server component. Once the call has been processed, we want, of course, to send a reply back to the client. This can be done with the `reply` operation. For example, a successful `update` invocation for a previously unknown key could be answered like this:

```
const charstring c_noPreviousValue := "";
pt.reply( update:{-,c_noPreviousValue} );
```

As you would expect, the specified port must be mapped or connected, must be of procedure kind and must list the given signature among its `in` signatures. The template used in the `reply` operation must specify the signature for which the reply is sent and give fully defined values for each `out` or `inout` parameter of the signature; `in` parameters may be omitted from the list using the '−' symbol. Therefore, in the previous example, no value is given for the `key` parameter, because it has been declared as an `in` parameter.

It is also possible, and indeed necessary, to specify a return value if the signature defines a return type. This is done using the `value` keyword as in:

```
pt.reply( lookup:{-} value "secret" );
```

It is important to mention that the return value has to be specified as a *value* and not, like the parameter values, as a template. This means that it is not possible to use a template reference in place of the return value or to prefix the returned value with its type. The following code is syntactically incorrect:

```
pt.reply( lookup:{-} value charstring:"" ); // syntax error!
```

If it is wished to use an explicit template to specify the return value to send to the client, the `valueof` operation can be used to turn it into genuine value first. This operation will be discussed in more detail in Chapter 10.

It is permitted, although unusual, to use the `reply` operation without a preceding `getcall` statement, that is to answer calls that have not been made. This could, for example be used to test your SUT's behaviour if more than one reply is received for a single remote procedure invocation. Depending on the RPC mechanism, that is used to communicate with the SUT, it may not be possible to actually dispatch such orphan replies to the SUT, but it is not the TTCN-3 system that prevents you from trying.

6.6.3 The `raise` Operation

Errors that occur during the execution of a remote procedure are often signalled to the client using exceptions. In TTCN-3, such an exception is generated with the `raise` operation. For example, failure to look up a requested key could yield the following exception.

```
pt.raise( lookup, NotFound:{} );
```

The specified port must be mapped or connected and have an underlying port type, that is of `procedure` kind and lists the given signature among its `in` signatures. Additionally, the signature, for which the exception is generated, has to be specified together with a fully defined implicit or explicit template for the exception's type. Note that, similar to the `catch` operation, only the signature name and not a signature template must be given. No parameters are passed back in the exception case and hence this information suffices.

Like the `reply` operation, the `raise` operation can be used without a previous `getcall` operation. It depends on your underlying RPC mechanism, if such an orphan exception can actually be passed to the SUT.

6.7 Addressing

We have now covered all communication operations for procedure-based communication, but we have so far ignored all aspects of addressing, like one-to-many port configurations, recipient specification or sender restrictions. Addressing aspects are very similar for all procedure-based communication operations. We will now discuss these aspects in more detail, see also Section 5.5, which covers addressing for message-based communication.

Our modelling of the communication between the directory server and its client has so far only been appropriate for a single server and a single client that communicate via one pair of connected ports of type `DirectoryServer` and `DirectoryClient`, respectively.

This approach is appropriate as long as the number of clients/servers is small and there is a fixed (and small) upper bound on the number of other peers that it is communicating with at any point in time. If a server should be capable of serving an arbitrary number of clients, then we need a mechanism to unambiguously identify the communication partners. This can be done by adding `from` and `to` clauses to the communication operations, as is done for message-based communication. It is also possible to determine the originator of procedure-based communication events using the sender redirection, as we have also seen for message-based communication.

When communication takes place between parallel test components, the TTCN-3 test system is responsible for assigning component addresses to each communication partner. When communication is with the SUT via the TSI, then it is up to the TTCN-3 developer to use an appropriate addressing mechanism. If your underlying RPC mechanism is, for example CORBA, then CORBA's interchangeable object references (IORs) could be used as addresses. The code in Table 6.21 could be used in a directory server implementation that supports only the `lookup` method and that serves an arbitrary number of clients over a single `DirectoryServer` port `pt`.

You can see how the sender of the incoming call is redirected to the variable `v_theClient` and subsequently used to direct the reply back to the calling client. Of course, similar mechanisms for sender redirection exist for the `getreply` and `catch` operations, as well as the possibility to address the receiving party in case of the `call` and `raise` operations.

It is important to note that this redirection of a variable of type `address` is only allowed if communication takes place via ports that are mapped to a TSI port – when communicating between parallel test components, a variable of suitable component type must be used to store the communication partner. So the example from Table 6.21 works only as long as the test component port `pt` is mapped to a TSI port. Of course, this becomes problematic when one and the same component shall offer the communication to both entities in the SUT and inside the test system. In this case, two different ports with duplicated code that differs only in the type of variables used for sender redirection may be used. Alternatively, it may be possible to provide some form of loop back mechanism

Table 6.21 Using addresses in procedure-based communication

```
var address v_theClient;
var charstring v_theKey, v_theValue;

// main server loop
while ( true ) {
  alt {
    [] pt.getcall( lookup:{?} ) -> param ( v_theKey ) sender v_theClient {
        if ( f_inSession( v_theClient ) ) {
          if ( f_keyExists( v_theKey ) ) {
            pt.reply( lookup:{-} value f_lookupKey( v_theKey ) )
              to v_theClient;
          }
          else {
            pt.raise( lookup, NotFound:{} ) to v_theClient;
          }
        }
        else {
          pt.raise( lookup, SessionExpired:{} ) to v_theClient;
        }
      }
  }
}
```

in the test system adapter for inter-component communication, which therefore can be then treated as ordinary communication via the TSI. More information on test system adaptation is provided in Chapter 12.

It is also possible to restrict the accepted calls, replies or caught exceptions using a from clause. For example, the directory server may have an administrative shutdown method, which may only be used by the directory manager to terminate the directory server. Calls to the shutdown procedure should only be accepted from the well-known directory manager to prevent users from accidentally or maliciously shutting down the directory server. This could be accomplished as shown in Table 6.22, where we assume that the variable v_myDirectoryManager contains the address of the manager from which we are prepared to accept a shutdown call. Of course, from clauses may also be used with the getreply and catch operation.

Finally, we can now come back to the DirectoryManager interface from the IDL example in Table 6.1 and show how address values can be used to return handles to be used in further communication operations. With our previous discussion of the use of the address type, we are now able to give an appropriate mapping of this interface in

Table 6.22 Accepting calls from specified components

```
alt {
  [] pt.getcall( shutdown:{} ) from v_myDirectoryManager {
      stop;
    }
  [] pt.getcall( shutdown:{} ) {
      // non-authorized shutdown attempt
      setverdict( fail );
    }
}
```

Table 6.23 TTCN-3 definitions for the DirectoryManager interface

```
type address DirectoryObject;

signature login( in charstring username, in charstring password,
                 out Capabilities capabilities )
return DirectoryObject
exception ( Rejected );

type enumerated Capabilities { e_reader, e_readerwriter };

type port DirectoryManagerServer procedure {
  in login
}
type port DirectoryManagerClient procedure {
  out login
}
```

Table 6.24 Logging into the directory service

```
// pt_manager is mapped to a port of type DirectoryManagerClient
// pt_directory is mapped to a port of type DirectoryClient (see Table 6.3)

var address v_theDirectory := null;

pt_manager.call( login:{username := "DS-user",
                        password := "secret"}, 5.0 ) {
  [] pt_manager.getreply( login:? ) -> value v_theDirectory {
        // we are logged in - test if we can lookup our password
        pt_directory.call( lookup:{"password of DS-user"}, 2.0 )
                    to v_theDirectory {
          [] pt_directory.getreply( lookup:? value "secret" ) {
               setverdict( pass );
             }
          [] pt_directory.getreply( lookup:? ) {
               setverdict( fail );
             }
          [] pt_directory.catch( timeout ) {
               setverdict( fail );
             }
        }
     }
  [] pt_manager.catch( timeout ) {
        setverdict( fail );
     }
}
```

TTCN-3 as shown in Table 6.23. You can see that the `address` type is used to define the return value of the `login` method.

Note again that these definitions will not work if used for communication between two parallel test components because values of type `address` can only be used to address communication from or towards the SUT. Assuming that the directory service is located in the SUT, a basic test that checks if we can log into the service as user `DS-user` and then look up our own password would look as shown in Table 6.24. We are using the address, that is returned by the initial `login` as a handle for the further communication with the SUT. It is up to the test system adapter to use this information to route our call to the appropriate object in the SUT. This is clear also from the fact that we have not changed the test system configuration on the basis of the address information that we have received from the `login` procedure. One port is dedicated to invocations from `Directory` interface, regardless of the address of the called server.

6.8   Summary

This concludes our discussion of procedure-based communication. We have shown how signatures can be used to map remotely callable procedures to TTCN-3, and how TTCN-3's procedure-based communication operations can be used for both client- and server-side communication. We have seen that the prevailing mode of procedure-based communication in TTCN-3 is that of synchronous calls, where a client waits for the result of his call to return; it is also possible to call functions in an asynchronous manner.

Finally, we have discussed how addressing can be used to emulate object references, which are commonly encountered in the realm of distributed systems, for example CORBA-based systems. We will come back in Section 9.5.2 to the issue of testing systems with procedure-based communication in our discussion of importing IDL types into TTCN-3 test system.

7

Modular TTCN-3

In previous chapters, we have focused on the development of more or less isolated test cases. Now in this chapter we take a step back and introduce one aspect of TTCN-3 that concerns the development of collections of test cases or test suites, namely, modularity. In addition to concurrency and testing-related constructs, strong support for modularity is probably one of the key features of TTCN-3.

As you have learned in our initial chapters, all TTCN-3 code exists within modules. At least one module is required, which may contain all your TTCN-3 code, but from a language point of view, there is no limit on the number of modules that you may use for structuring our code. In this chapter, we will show how to work with multiple modules, how to import TTCN-3 definitions from one module to another, and how in practice to structure a test suite into several modules.

Modularity and modularisation of TTCN-3 code are important because they can provide the key to a successful testing project. When several TTCN-3 developers work together, modularisation allows easier distribution of code development and maintenance. By using a sensible structuring of your code, existing code can easily be located, which improves reuse and thus reduces the amount of code to write and maintain. Modularisation can also help decrease the turnaround times during the development of large test suites: smart TTCN-3 tools will only re-process those modules that have been changed since the last processing. Finally, modularisation is also the way in which extensibility of TTCN-3 towards other languages is achieved.

Tightly coupled with modularisation is the notion of importing, that is using definitions from one module within another module. We will discuss how this is expressed in TTCN-3.

Groups provide an additional way to structure code within a single module and allow for selective import of related definitions. They will be discussed briefly in this chapter.

While discussing modules and modularity, we will also treat TTCN-3 module parameters. They allow the specification of TTCN-3 values outside of your TTCN-3 code. These are external values in the sense that they may be changed at execution time without the necessity to re-process the TTCN-3 code. Module parameters are commonly used to handle parameters that are specific to the System Under Test (SUT), for example the

An Introduction to TTCN-3, Second Edition.
Colin Willcock, Thomas Deiß, Stephan Tobies, Stefan Keil, Federico Engler and Stephan Schulz.
© 2011 John Wiley & Sons, Ltd. Published 2011 by John Wiley & Sons, Ltd.

IP address for our Domain Name System (DNS) server, that is to be tested. Module parameters allow the design of test suites independent of a specific SUT instance. They allow using the same test system executable in different environments, for example against a local DNS server in our own laboratory or somewhere out in the field.

In this chapter, we will also introduce TTCN-3 attributes. Attributes allow the specification of meta-information for TTCN-3 definitions inside the textual TTCN-3 core notation. Such meta-information is not accessible from within TTCN-3 but may convey important information to be used by other parts of a test system, for example codec implementations or graphical TTCN-3 editors.

7.1 Modules

In this section, we cover the basics of TTCN-3 module definition and also propose a structure to modularise test suites. This structure is particularly useful for protocol testing purposes but may be equally applicable in other testing contexts.

7.1.1 Definition of a Module

The transfer syntax for TTCN-3 modules is UCS Transformation Format 8 (UTF-8). Each character of a module definition shall be individually encoded and decoded according to the UTF-8 rules as defined in annex R of ISO/IEC 10646 [33].

A module definition starts with the `module` keyword followed by an identifier, that is the module's name. When structuring your test suite into several modules, each module must have a unique name. The module body is delimited by curly brackets and may contain an arbitrary number of definitions and at most one control part. Placing any TTCN-3 definitions outside of a module is not allowed.

In theory, TTCN-3 does not place any restriction on the order of definitions within a module (except that the control part, if present, comes last). In particular, it is not necessary to make module-level definition before they are referenced. This exception of TTCN-3's declare-before-use policy makes it possible to place definitions within a module without any ordering constraints and to define mutually recursive type structures or functions without explicit forward references.

It is advisable, though, to use common sense when structuring the TTCN-3 code inside a module. Placing related definitions close together will improve the readability of the code but makes locating of certain definitions more complicated. On the other hand, sorting definitions by name and/or kind will help to find information quickly, but will spread related information over several places of your code. As a rule of thumb, the larger your individual modules become, the more useful a stringent ordering, for example by definition kind and lexicographic order of identifiers, will be. Groups, which will be described further down, may also be used to give additional structure to your code. Table 7.1 gives an example of a module that has been largely structured using a 'declare-before-use' strategy. Such a strategy is particularly useful for small modules that can be easily read in their entirety.

When present, a control part must be the last definition in your module. A control part is responsible for the selection and execution of test cases via the `execute` statement.

Table 7.1 Example module definitions for DNS protocol types

```
module DnsProtocolTypes {
  const integer c_defaultDnsPort   := 53;
  const integer c_unsignedShortMax := 65535;

  type charstring   Answer;
  type integer      Identification( 0 .. c_unsignedShortMax );
  type enumerated   MessageKind { e_question, e_answer };
  type charstring   Question;

  type record DnsMessage {
    Identification  identification,
    MessageKind     messageKind,
    Question        question,
    Answer          answer optional
  }

  type octetstring RawDnsMessage;
}
```

This statement has been described in Section 4.3.3. You can think of the control part as your module's main function. This main function is invoked upon the test suite's execution. Multiple modules may specify their individual control parts, and it is up to the TTCN-3 tool or an external test control entity to select which control part to run during test suite execution.

Although the TTCN-3 standard provides no restrictions on the number of modules per source file, you are well advised to place each module in a single file and use the module name also as the file name. This will greatly simplify locating of source files for referenced modules.

7.1.2 Modularisation of TTCN-3 Test Suites

Once a TTCN-3 test suite reaches a certain complexity, modularisation will be mandatory to retain readability and maintainability. It also helps to foster reuse of your code, if definitions are easily locatable by different TTCN-3 programmers. The TTCN-3 core language standard itself neither mandates nor suggests any guidelines on how a test suite should be modularised. Although your approach to modularisation will depend on your specific application area of TTCN-3 and the scale of your test development effort, we have found that a few simple guidelines will help achieve a first coarse structure that may then be refined depending on your more specific requirements. We will outline this structure in Section 13.3.

7.2 Group Definitions

Within a module, TTCN-3 definitions can be further structured using groups. In TTCN-3, groups have little logical significance, for example they do not form their own separate

Table 7.2   DNS protocol type module definitions with groups

```
module DnsProtocolTypes {
  group consts {
    group basic {
      const integer c_defaultDnsPort   := 53;
      const integer c_unsignedShortMax := 65535;
    }
  }

  group types {
    // the same group identifier may be reused within a different group
    group basic {
      type charstring   Answer;
      type integer      Identification( 0 .. c_unsignedShortMax );
      type charstring   Question;
      type octetstring  RawDnsMessage;
    }

    group structured {
      type enumerated   MessageKind { e_question, e_answer };

      type record DnsMessage {
        Identification  identification,
        MessageKind     messageKind,
        Question        question,
        Answer          answer optional
      }
    }
  }
}
```

scopes that could be used to control the visibility of definitions with a module. One
significance of groups is that they can be used for more selective importing of definitions
from module to module. Also, they can be used to add additional structuring of code in a
way that can be recognised by TTCN-3 programmers for easier navigation within modules.

A group definition is syntactically similar to a module definition, with the `group`
keyword followed by a group identifier and the contained definitions between curly braces.
Group definitions may be freely nested as long as there are no name clashes between sub-
group siblings within a module group. This is shown in Table 7.2, where `basic` is used
as a group name both in the group `consts` and `types`. It would not be allowed to have
another group `types` within the module, or to have another group `basic` within the group
`consts`. Note that using groups in such a small example may look artificial and contrived
but that grouping can be a valuable method to add structure to large modules, in particular
when your TTCN-3 editor allows navigating your TTCN-3 code using these groups.

7.3   Importing

With the possibility of modularising test suites into separate modules comes the need to
use definitions from one module in another module. This is achieved in TTCN-3 using

Table 7.3 Importing all definitions from another module

```
module DnsProtocolTemplates {
  import from DnsProtocolTypes all;

  template DnsMessage a_DnsQuestion( Identification p_id,
                                     Question       p_question ) := {
    identification := p_id,
    messageType    := e_question,
    question       := p_question,
    answer         := omit
  }
}
```

`import` definitions. What TTCN-3 adds to what is often available in other programming languages is a rich means for selective imports. Restrictions of imports are of interest as they can prevent name clashes, help establish clear interfaces between modules, and potentially reduce processing time and memory footprint of your test system.

Definitions from a module are made 'visible' in another module by explicitly importing them using an `import` definition. So, to define templates on the basis of the type definitions from the module `DnsProtocolTypes`, we import them as shown in Table 7.3. As you can see, an import definition starts with the keywords `import from` followed by the name of the module that we are importing from, followed by a specification of the definitions that we wish to import from that module. In the simplest case, `all` can be used to import all the definitions from a specified module. In the case of our example, this makes all module-level definitions from `DnsProtocolTypes` known in the module `DnsProtocolTemplates` and they may be used as if they were defined in the latter module. So, for example the types `DnsMessage`, `Identification` and `Question` from `DnsProtocolTypes` are used in the definition of a_DnsQuestion. It is possible to specify `import` definitions everywhere on the module level, though most commonly they are placed at the beginning of a module, as shown in Table 7.3.

7.3.1 Visibility of TTCN-3 Definitions

TTCN-3 has three types of visibility for definitions made at the top-level in the module definitions part. If no visibility type is specified with the definition it is `public` by default, which means it can be imported by any other module. Only for groups and imported definitions special rules apply as is described further below in this chapter. The other two types of visibility are `friend` and `private`.

A definition that has the visibility `private` can only be used locally in that specific module and not be imported by any other module. By default all imported definitions in a module have the visibility `private` and cannot be imported again from other modules.

A module can declare other modules to be friends as shown in Table 7.4.

Modules that have a friend-relationship can import each other's definitions that have the visibility `friend`. For modules without the friend-relationship definitions with the friend visibility cannot be imported and have the same implications as private definitions.

Table 7.4 Declaring other modules to be friends

```
module StarterM {
        friend module StopperM;
        friend type boolean IsStarted;
        // type IStarted is visible for import from module StopperM
}
```

Table 7.5 Example of mixed visibility

```
module StarterM {
        friend module StopperM;

        friend type boolean IsStarted;
        // IsStarted is visible from module StopperM

        public group allIneed{
                public  type integer    StartCount;
                friend  type boolean    HasBeenCounted;
                private type charstring OnlyForStarters;
                // OnlyForStarters is not visible from module StopperM
        }// group allIneed is visible from module StopperM
}
```

A group combines definitions that can have different types of visibility. A group definition itself is always public, no other visibility type can be specified for a group. So what happens if a public group is imported into another module but its members have mixed visibility types? Here the same rules apply that are valid for importing those definitions directly: private definitions will not be imported at all even though they are part of the group as they are local to the module they are defined in. Friend definitions are only imported if the two modules have an existing friend-relationship or otherwise they are invisible. So a group with members of mixed visibility will contain different definitions depending on the relationship between modules and the content of the group. See Table 7.5.

Imported definitions have by default private visibility inside the importing module unless a different visibility is explicitly stated before the import statement. This makes it possible to import definitions into a module that were not imported directly by the module itself but by another module. For example, for a group of modules with friend relationship it could be reasonable to have one module importing definitions that are needed in the other modules and providing them to those with the friend visibility, as shown in Table 7.6.

In the following section, we will discuss this import mechanism in more detail.

7.3.2 About Transitivity of Imports and Cyclic Imports

Importing definitions from a module works by default in a non-transitive manner: only the local definitions of that module can be imported directly, but not the definitions that it

Table 7.6 Example of friend visibility

```
module StarterM {
      friend module StopperM;
      friend type boolean IsStarted;

      friend import from TemplateCollection template all;
      // module StopperM can see all imported templates from
      // module TemplateCollection
      import from StarterTypes type all;
      // module StopperM cannot see the types from module
      // StarterTypes
      public import from PublicTypes type all;
      // any module importing definitions from module StarterM
      // can see the type definitions from module PublicTypes
}
```

imports. Such definitions will have to be imported separately from the defining module. For example, after importing all definitions from the module DnsProtocolTemplates, none of the types or constants from the module DnsProtocolTypes will be usable in the module DnsTestCases although they are imported into DnsProtocolTemplates. Only after explicitly importing from DnsProtocolTypes will it be possible to use the definitions from that module, for example the type Identification.

```
module DnsTestCases {
  import from DnsProtocolTemplates all;
  import from DnsProtocolTypes all;

  const Identification c_defaultId := 12345;
}
```

The updated visibility rules make importing explicitly transitive by also making imported definitions accessible for further imports. Note that having cyclic import relations among modules is allowed.

7.3.3 Restricting the Import of TTCN-3 Definitions

As visibility rules apply for importing definitions, importing all definitions will only import those that are visible from the importing module. So a part of the restrictions are set by the visibility stated for each top-level definition making it explicitly public, restricting access to friend-related modules or prohibiting access by making a definition private. In its simplest form, the import statement does not restrict the imported definitions at all. This is expressed with the keyword all in the import definition:

```
import from DnsProtocolTypes all;
```

In this case, all definitions defined in the module DNSProtocolTypes are imported. Definitions imported to the module DnsProtocolTypes are not imported.

In practice, there are good reasons **not** to use this unconstrained form of import and to restrict the definitions imported into a module. Firstly, restrictive imports lead to smaller 'interfaces' between modules, which improves maintainability. When importing all definitions, this conveys no information about which of the imported definitions are really used in the importing module that may be substantial in size. So it becomes hard to estimate the impact that a change to the imported definitions will have on the importing module. Secondly, unrestricted imports increase the amount of work that TTCN-3 tools have to perform in the processing of a module prior to the execution of the test system.

7.3.3.1 Restriction by Kind

TTCN-3 allows restricting imports on two levels: firstly, it is possible to restrict the import to certain kinds of definitions; secondly, it is possible to import specific definitions identified by their name. So, to make it explicit that we are only importing types and constants from DnsProtocolTypes, the following import definition could be used.

```
import from DnsProtocolTypes {
  type  all;
  const all;
}
```

The available definition kinds are group, testcase, function (which includes external functions), altstep, template, type, signature, const and modulepar (yet to be introduced in this chapter). Note that timer and variable declarations cannot be imported because they can never be declared on the module level. Also a control part can never be imported.

7.3.3.2 Restriction by Name

When a more fine-grained control over the imported definitions is needed, then it is possible to name individual definitions to be imported:

```
import from DnsProtocolTypes {
  const c_defaultDnsPort;
  type  DnsMessage, RawDnsMessage;
  type  Identification;
}
```

Also, mixing both forms of import restrictions is possible:

```
import from DnsProtocolTypes {
  const c_defaultDnsPort;
  type  all;
}
```

7.3.3.3 Importing Groups

When importing a group or sub-group, all definitions from that group are imported depending on the visibility for each group-member (see Section 7.3.1). Note that it is possible to import sub-groups directly, that is without their surrounding parent group, as long as it has a unique name. Coming back to the module `DnsProtocolTypes` from Table 7.2, we can observe the following group structure.

```
module DnsProtocolTypes {
  group consts {
    group basic { /* ... */ }
  }

  group types {
    group basic { /* ... */ }
    group structured { /* ... */ }
  }
}
```

Here, it is possible to import the group `structured` directly, even though it is a sub-group:

```
import from DnsProtocolTypes {  group structured }
```

It is **not** possible to import the group `basic` in this manner – there are two sub-groups with this name. If one of these groups shall be imported, it needs to be explicitly qualified with its parent group's name:

```
import from DnsProtocolTypes { group basic } // ERROR: not unique
import from DnsProtocolTypes { group consts.basic } // OK
```

7.3.3.4 Expressing Exceptions When Importing All

Finally, the TTCN-3 language also offers the possibility of importing all definitions (of a certain module, kind or group) *except* for an explicitly excluded list of definitions. This is indicated using the `except` keyword. For import definitions that import all definitions of a certain kind, the definitions to be excluded are simply listed in the import:

```
import from DnsProtocolTypes {
  type all except Identification, RawDnsMessage;
  const all except c_unsignedshortMax;
}
```

When importing definitions of different kinds, for example all definitions from a module, or a group, it is also possible to exclude specific definition kinds from the import.

```
import from DnsProtocolTypes all except { const all }

import from DnsProtocolTypes {
  group types.basic except {
    type RawDnsMessage;
    template all;              // OK, there are none there anyway
  }
}
```

One possible application area for this form of restriction is to remove name clashes between imported identifiers, which will be discussed in our next section.

7.3.4 Module Prefixing of Imported Definitions

When imported definitions are used within the importing module, it is possible to indicate the origin of the definition by prefixing the imported identifier with the name of module where it has been imported from, separated by a dot. Module identifiers may be quite long though, and a (desirable) high level of modularisation increases the use of imported definitions. This means that the effect of always prefixing every imported definition identifier with its module name may lead to rather unreadable TTCN-3 code.

There are cases when prefixes are useful or even necessary, in particular, when there are clashes between an imported identifier and a local identifier, or between two or more imported identifiers. When there is a clash between an imported identifier and a local identifier, the local identifier shadows the imported one. Only the qualification of the identifier with the module name allows the imported definition to be accessed. This is shown in Table 7.7.

When such a name clash occurs between two imported modules and there is no local definition with the same name, no precedence is given to any of the definitions and the only legal way to use the clashing identifier in the importing module is in the qualified form:

```
import from DnsProtocolTypes { const c_defaultDnsPort };
import from DnsTestSystem    { const c_defaultDnsPort };

// ERROR - ambiguous identifier
const integer c_p1 := c_defaultDnsPort;
// OK - assigns 53
const integer c_p2 := DnsProtocolTypes.c_defaultDnsPort;
// OK - assigns 54
const integer c_p3 := DnsTestSystem.c_defaultDnsPort;
```

Alternatively, it is possible to exclude clashing identifiers explicitly from import definitions by using the **except** restriction.

```
import from DnsProtocolTypes all except {
                            const c_defaultDnsPort };
```

Table 7.7 Resolving a name clash between local and imported definitions

```
module DnsTestSystem {
  import from DnsProtocolTypes { const c_defaultDnsPort };
  const integer c_defaultDnsPort := 54; // shadows the imported const

  // assigns 53 (imported)
  const integer c_p1 := DnsProtocolTypes.c_defaultDnsPort;
  // assigns 54 (local)
  const integer c_p2 := c_defaultDnsPort;
}
```

7.3.5 Transitive Import

When importing a module only the definitions defined in that module can actually be imported, but not the definitions imported into that module. This is neither possible with importing all definitions, nor with importing specific definitions. By default an `import` statement has `private` visibility, but this can be changed to `public` or `friend`. In this case the visible `import` statements of a module can again be imported to another module. To import all visible import definitions of the module B, the statement `import from B { import all }` is used. It is not possible to import individual `import` statements.

7.3.6 Importing from Other Languages

We have already stated that TTCN-3 was designed to be extensible towards other programming languages and type systems. The `import` statement plays the key role in the extension of TTCN-3 code to embrace other languages. For example, ASN.1 type definitions can be used within TTCN-3 after importing them into TTCN-3 modules, while specifying that the imported information is not TTCN-3 but instead ASN.1:

```
import from DNS language "ASN.1:2002" all;
```

More on the specifics of how foreign languages – for example, type definitions such as Abstract Syntax Notation 1 (ASN.1) and Interface Definition Language (IDL) – are integrated into TTCN-3 can be found in Section 9.5.

7.4 Module Parameters

TTCN-3 has advanced communication primitives for communication with the SUT but no mechanisms for user-interaction. TTCN-3's prime applicability lies in the *automatic* execution of test suites against the SUT, so a reliance on user input would be a hindrance rather than an advantage. On the other hand, there is the necessity to provide certain parameters to a test suite, so that it can successfully execute in different environments. Details, for example the actual SUT address, the selection of a specific test case execution and so on, are examples of such parameters, which may be provided prior to the execution of a test suite.

Table 7.8 Defining module parameters with and without default values

```
module DnsParameters {

  import from DnsProtocolTypes { const c_defaultDnsPort }

  modulepar charstring mp_sutIpAddress, mp_localAddress;

  modulepar integer mp_localPort := 1059;
  modulepar integer mp_sutPort    := c_defaultDnsPort;
}
```

These parameters are called module parameters in TTCN-3. Module parameters can be used to provide external parameters to a TTCN-3 test suite at execution time, that is without the need to re-process the TTCN-3 code for each parameter value modification. In many ways, module parameters work like constants, which can be overwritten externally by the test system user upon test system execution. After that, their value may not be modified during the execution of the test system. The change of a module parameter will be reported as an error by a TTCN-3 system.

By providing different values for different runs, specific address information for connections to the SUT or some simple form of test case selection can be achieved.

Module parameters are declared on the module level, using the `modulepar` keyword followed by one or more module parameters between curly brackets. Each module parameter has to be declared with a type and may optionally specify a default value. Some examples are shown in Table 7.8.

How the actual values of module parameters are specified externally will depend on your TTCN-3 tool. Usually this is done via command line parameters or configuration files. The TTCN-3 Control Interface (TCI) standard [13] also specifies an option via which module parameters can be externally provided. See Chapter 13 for more information. If no actual value for a module parameter can be found at runtime, the default value will be used. If no default value has been specified either, then the first access of the module parameter value will cause a test case error.

Although writing to module parameters is not allowed, they are treated differently from constants: using them in contexts in which only constant values are allowed, such as in the definition of subtyping constraints or array boundaries is not permitted.

```
import from DnsParameters { modulepar all }

// ERROR when used to init const
const integer c_localPort    := mp_localPort;
// OK when used in templates
template integer a_localPort := mp_localPort;
```

7.5 Attributes

TTCN-3 allows the specification of meta-information, that is information *about* your definitions, inside the TTCN-3 core notation. This information is inaccessible from within your TTCN-3 code. It cannot be read or written by any statement. Instead, it carries information that can be used by test system entities external to your test suite, like codec implementations or graphical TTCN-3 editors.

This information is specified in the form of attributes, which may be assigned to an (import) definition, group or module using the keyword with:

```
type integer Identification ( 0 .. 65535 )
        with { variant "unsigned 16 bit" }
```

TTCN-3 allows attributes of five different kinds:

- display for the specification of information, that is related to different TTCN-3 presentation formats (other than the core notation) and which is relevant for graphical TTCN-3 editing tools;
- encode for the specification of information that will be used by the codec implementation (like "BER:1997" to select encoding of data using ASN.1's basic encoding rules);
- variant for the specification of information that selects a certain available variation within the selected encoding (like "unsigned 16 bit" for integer values or "UTF-8" for character strings);
- extension for user- or tool-specific purposes;
- optional for indicating whether optional fields are set to be absent by leaving them out from definition or by setting them to omit, see Section 8.4.2.1.

Of all these attributes only encode, variant and optional are of relevance to a TTCN-3 developer. The attributes display and extension are tool-related and will not be discussed here any further.

Attributes themselves are always given as a character string, and a number of values have been given standardised meanings. However, it should be noted that implementation of these standard strings may be tool dependant. Examples of these standardised strings are:

- "BER:1997" or "PER-BASIC-UNALIGNED:2002", to select specific, standardised encodings for ASN.1 defined data types.
- "8 bit", "IEEE754 extended float" or "IDL:fixed FORMAL/01-12-01 v.2.6" to request certain forms of encoding variant for integer and float values.

For a complete list of the standardised encode and variant attribute strings, refer to the TTCN-3 core language standard [8].

7.5.1 Accessing Attribute Values

We have already said that attributes are not accessible from within your TTCN-3 code, so how *can* the attributes be accessed? For the display and extension attributes, this is a tool-specific issue. For the encode and variant attributes, the actual attribute values for the types are visible at the interface that the TTCN-3 test system offers towards

codec implementations. This may either be a tool-specific, proprietary interface or the CD
interface, see Chapter 12, from the TCI standard [13]. Through these interfaces, the values
can be read and be used to control the behaviour of the codec implementation.

7.5.2 Scoping of Attributes

In their simplest form, attributes are simply attached to a single definition:

```
type integer Identification ( 0 .. 65535 )
        with { variant "unsigned 16 bit" }
```

It is also possible to assign `encode` or `variant` attributes to (groups of) specific
fields of structured types or indeed embedded sub fields as shown in Table 7.9.

Note that the assignment of the `"unsigned 16 bit"` attribute to the `identifi-
cation` field of `DnsMessage` is actually redundant because it has already been assigned
to the field's type `Identification`. This assignment is not shadowed by the `"32 bit
padded"` attribute, that is assigned to the structured type `DnsType`.

Looking at a larger scale, it is possible to attach attributes to whole groups of definitions
or even whole modules as shown in Table 7.10.

With the possibility of specifying `encode` or `variant` in several, possibly nested
contexts, there comes the issue of which attributes actually apply for a specific
definition. The rule here is that the innermost attribute specification takes precedence.
So, in our example above, the types from the group `types.basic` will carry the
`encode` attribute `"TrivialEncoding"`, while the types from `types.structured`
carry the attribute `"DnsEncoding"` and all other type definitions outside these
groups carry the attribute `"NoEncoding"`.

Table 7.9 Assigning `encode` or `variant` attributes

```
type record ExtraType{
   integer      subfield1,
   integer      subfield2
}

type record DnsMessage {
   Identification  identification,
   MessageKind     messageKind,
   Question        question,
   Answer          answer optional,
   ExtraType       extraField optional
}
with {
   variant ( identification )          "unsigned 16 bit";
   variant ( messageKind )             "1 bit flag";
   variant ( question, answer )        "7 bit ASCII with 16 bit length";
   variant ( extraField.subfield1)     "unsigned 16 bit";
   variant                             "32 bit padded";
}
```

Table 7.10 Assigning attributes to whole groups or modules

```
module DnsProtocolTypes {
  group types {
    group basic       { /* ... */ } with { encode "TrivialEncoding" }
    group structured { /* ... */ } with { encode "DnsEncoding" }
  }

  /* ... */

} with { encode "NoEncoding" }
```

There is also the possibility of overriding the attributes assigned to an inner scope from an outer scope using the keyword `override`, which can be used to force all types from a scope to a single attribute. The combination of normal attribute assignment with attribute assignments using `override`, when used excessively, will make it difficult for a reader of the TTCN-3 code to figure out which attributes will actually be applied to the different definitions. Hence, `override` should be used carefully.

7.5.3 Assigning Attributes to Imported Definitions

Usually, a definition retains its attribute when it is imported into another module. It is possible, though, to assign and/or change attributes when importing definitions. For this, the attributes are simply attached to the import definition as shown in Table 7.11.

When the defining module already assigns attributes to the imported definitions, then the attributes assigned in the import definition are treated as if they would be assigned to an additional scope enclosing the imported definitions. For our example, this would mean that the assignment of the `"SpecialDnsEncoding"` attribute would be ineffective because these types are already assigned the `encode` attributes `"TrivialEncoding"` or `"DnsEncoding"` depending on the group that they were defined in. This could be changed with an `override` directive:

```
import from DnsProtocolTypes { type all }
  with { encode override ( DnsMessage ) "SpecialDnsEncoding"
}
```

7.5.4 Using Attributes to Define Encodings

The use of attribute values is not restricted to the standardised strings shown in the examples so far. For example, it is possible to give the complete encoding information for the text-based protocol, like the Session Initiation Protocol (SIP), in the form of encoding attributes. Table 7.12 shows one example where encoding attributes specify

Table 7.11 Assigning attributes when importing

```
import from DnsProtocolTypes { type all } with
                           { encode "SpecialDnsEncoding"; }
```

Table 7.12   Example use of encoding attributes for textual encoding

```
type record HostPort {
  charstring    host,
  unsignedshort portNumber optional
} with {
  encode ( portNumber ) ":_";
};

type record UserInfo {
  charstring name,
  charstring password optional
} with {
  encode ( password ) ":_";
};

type record SipUri {
  UserInfo    userInfo optional,
  HostPort    hostPort,
  charstring uriParams optional,
  charstring headers optional
} with {
  encode "sip:_";
  encode ( userInfo ) "_@";
  encode ( uriParams ) ";_";
  encode ( headers ) "?_";
};

const SipUri c_uriSip4Alice := {
  userName   := { "alice", omit },
  hostPort   := { "atlanta.com", 5060 },
  uriParams  := omit,
  headers    := omit
};
```

the encoding of a SIP Uniform Resource Identifier (URI). In this example, each encoding attribute is specified using an encoded string value. In the specific case of the HostPort field portNumber, this encoded string expresses that a colon must be added by an encoder *prior* to the encoded portNumber value. If a string follows after the underscore, as in the case of the userInfo field of the SipUri type, it is to be appended to the encoded value. The result of encoding the constant value c_sipUri4Alice would therefore yield "sip:alice@atlanta:5060". For more details on the use of encoding attributes for defining a textual encoding, refer to [35].

7.6 Summary

In this chapter, we have studied the aspects of TTCN-3 that are concerned with the development of whole test suites rather than individual test cases. We have studied how test suites can be split into separate modules to improve maintainability. We have shown how definitions from one module can be reused in another module by importing them. In addition, we have looked at the different forms of the import statement that can be used to restrict which specific definitions shall be imported from one module into another, both visibility of definitions can be specified in the exporting module and the to be imported definitions can be defined in the importing module. Importing definitions into a module may lead to clashes between local and imported identifiers, which can be resolved by the qualification of identifiers with their module names.

Another concept that was introduced in this chapter was module parameters. We have seen how module parameters in TTCN-3 allow the specification of configuration values to allow for execution of a test suite in different environments without the necessity to change or re-process the test suite itself.

Finally, we have studied attributes, which allow the specification of meta-information associated with TTCN-3 definition. These attributes can be used to control external properties of the TTCN-3 code, like its rendering in graphical TTCN-3 editors, the way that values are encoded and decoded when sent between the test system and the SUT, or how absence of optional fields can be defined. We have seen that TTCN-3 allows the specification of attributes for single definitions, groups and full modules, and we have discussed the way that the actual attribute for a definition is selected as well as a means to override default attribute precedence.

8

TTCN-3 Data Types

Data types, their declarations and their usage form the core of any programming language and are often the starting point of programming language manuals and tutorials. In this tutorial, we have chosen a different approach and introduced most features of the Testing and Test Control Notation Version 3 (TTCN-3) with only a minimal set of data types. The reason for this approach is that the range of data types in TTCN-3 is much larger than that usually found in typical programming languages. In this case, a complete coverage of data types would have been too distracting in the early stages of this book, when the aim was to enable the writing of TTCN-3 code as soon as possible. Now that the overview of the language has been given, it is time to present TTCN-3's data types in more depth.

The modelling in TTCN-3 of the application data or protocol data units (PDUs) for the testing domain is a crucial step in test system design. The most important point here is to get the right level of abstraction. TTCN-3 provides the possibility of creating a close correspondence between the data types of the system under test (SUT) and its representation in the test system. Finding the right mapping is not an easy task, but once a suitable representation has been found, many things in the test system will fall into place easily.

TTCN-3 features a large number of data types, many of which you will recognise from other programming languages (`boolean`, `integer`, `float`, `(universal) charstring`). Other data types are unique to TTCN-3 and reflect its usage as a test scripting language with a protocol testing background (`verdicttype`, `bitstring`, `hexstring`, `octetstring`, `default`). Additionally, TTCN-3 allows the definition of structured types (`enumerated`, `record`, `set`, `union`) and list types (`array`, `set of`, `record of`) from existing types. The reason that the range of built-in and user-defined types exceeds those from most other programming languages is that TTCN-3 has to be able to model the application data as closely and naturally as possible in a wide range of application domains.

Before we look at the TTCN-3 types, we will start by discussing subtyping. Subtyping refers to restricting types to only a subset of their possible value set. Many applications and protocols place such restrictions on the set of allowed inputs or information elements in

An Introduction to TTCN-3, Second Edition.
Colin Willcock, Thomas Deiß, Stephan Tobies, Stefan Keil, Federico Engler and Stephan Schulz.
© 2011 John Wiley & Sons, Ltd. Published 2011 by John Wiley & Sons, Ltd.

messages. By allowing for subtyping, TTCN-3 makes it possible to reflect such constraints directly in the test system. This makes it possible to prevent the sending of values to the SUT that cannot be legally encoded, for example because of field length restrictions.

After subtyping, we will then cover the entire TTCN-3 type system, both basic types and structured types. Each type will be introduced with its value notation, operators and some useful functions that TTCN-3 provides to process values of this kind.

Before we start to look at subtyping in more detail, we will spend some time introducing the Session Initiation Protocol (SIP) [26], which plays an important role in the area of Internet telephony. We will use this protocol to focus the examples that we use for the introduction of the various data types and will gradually build type definitions for SIP messages while exploring the various aspects of the TTCN-3 type system.

8.1   The Session Initiation Protocol

The SIP has been developed as a signalling protocol for creating, modifying and terminating sessions with one or more participants. These sessions include Internet telephone calls, multimedia distribution, conference calls and multi-player online games. SIP is a pure signalling protocol, that is it is not concerned with the transport of the session's content. Instead, it carries meta-information that describes the media content of the session, which is then used to set up the communication channels. During the establishment phase of a session, communication between the entities is carried out through proxy servers that perform important tasks such as locating the communication partners, forwarding of calls, creating billing information and so on.

To make things concrete, we give a few examples taken directly from RFC 3261, the Internet standard that defines SIP [26]. For example, to invite Bob to a call, Alice could send the message from Table 8.1 to her SIP proxy server.

The structure of this message is typical for a SIP request message: the first line identifies the request kind, the SIP destination of the request and the SIP version used. It is followed by a sequence of headers that, for example identify the SIP proxy at which the addressee is known to be reachable (Via), the specification of a limit on the number of times the messages may be forwarded between proxies (Max-Forwards), the originator of the request (From) and additional information about the payload of the message (Content-Type) with its length (Content-Length). In the given example, the payload contains the initial

Table 8.1   A SIP message inviting Bob to a call

```
INVITE sip:bob@biloxi.com SIP/2.0
Via: SIP/2.0/UDP pc33.atlanta.com
Max-Forwards: 70
To: Bob <sip:bob@biloxi.com>
From: Alice <sip:alice@atlanta.com>
Call-ID: a84b4c76e66710@pc33.atlanta.com
CSeq: 314159 INVITE
Content-Type: application/sdp
Content-Length: 142
<content not shown>
```

Table 8.2 Bob's SIP reply to the invitation

```
SIP/2.0 200 OK
Via: SIP/2.0/UDP bigbox3.site3.atlanta.com;
Via: SIP/2.0/UDP pc33.atlanta.com;'
To: Bob <sip:bob@biloxi.com>;tag=a6c85cf
From: Alice <sip:alice@atlanta.com>;tag=1928301774
Call-ID: a84b4c76e66710@pc33.atlanta.com
CSeq: 314159 INVITE
Content-Type: application/sdp
Content-Length: 131
<content not shown>
```

parameters for the voice call. This is one possible payload that may be carried by a SIP message: media session parameters specified in the Session Description Protocol (SDP).

After this request has finally reached Bob and Bob has agreed to establish the session with Alice, Bob might send the response from Table 8.2 back to Alice via his proxy.

This response consists of the SIP version followed by the response code (200 OK), followed by a sequence of headers. The `Via` headers specify the sequence of proxy servers that the message shall take on its way back to Alice. The From and To headers are repeated from the request, and again a description of the payload type and its size in octets are specified. In our example, the payload would contain a description of the kind of session that Bob is willing to participate in.

From these examples, you can see that each SIP message transmitted between the entities is a simple character string. This would make it possible to use character strings to represent these messages inside the TTCN-3 code. This is not the best approach though, because it makes inspection and construction of messages in the test cases unnecessarily complicated. These tasks become much simpler when the structure of SIP messages is reflected by their type representation in TTCN-3. Transformation between the unstructured string representation used to send SIP messages via the Internet, and the structured representation used to handle the messages within the TTCN-3 code is then handled by a suitable encoder/decoder implementation.

8.2 Subtyping

Subtyping, the definition of new types as a restriction of already defined or built-in parent types, is a well-known concept from other programming languages such as Ada [36], Modula-3 [37] or VHDL [38]. We have previously touched on this issue in the introductory Domain Name System (DNS) example in Chapter 2, when defining the type `Identification`, which was a subtype of `integer` restricted to those unsigned numbers representable with 16 binary digits. Since the TTCN-3 type system allows specification of arbitrarily large or small values – take the `integer` with its upper and lower infinite bounds as an example – subtyping allows us to restrict the values that can be set in a test which helps us avoid the specification of incorrect values that cannot be possibly represented in the real world. For a message field which has to be encoded in 1 bit, for example we should only be able to specify or send `integer` values 0 and 1. The use of

Table 8.3 Some examples for subtype violations

```
type integer unsignedSevenBit ( 0 .. 127 );

const unsignedSeventBit c_1K := 1024;        // ERROR - subtype violation
var integer              x := 512;
var unsignedSevenBit     y := x;             // ERROR - subtype violation

function f_int2ascii( in unsignedSevenBit p_x ) return char {
  // perform integer to ascii conversion and return value
};

var char illegalChar := f_int2ascii( 300 );  // ERROR - subtype violation
```

any other `integer` value would lead to a test case error when an encoder would attempt
to encode this field value for sending it to the SUT. For this reason, TTCN-3 supports a
wide range of subtype definitions:

- aliasing – giving a new name to an already defined type; this is available for all types;
- value lists – restricting a type to a list of admissible values; this is available for all
  types;
- value ranges – restricting an ordered type to a certain range; this is available for the
  `integer` and `float` types;
- field value constraints – restricting values of selected fields for structured types;
- character set restrictions – restricting the admissible characters in a character string
  type; this is available for the `charstring` and `universal charstring` types;
- length restrictions – restricting the number of elements in strings or list types; this is
  available for all string-like types and the `record of` and `set of` types.

Subtypes are enforced by the TTCN-3 system and are checked both during analysis
and during execution time. Subtype restrictions are violated whenever a value outside the
subtype's allowed value set is used where a value compatible with the subtype is required.
This may happen in variable assignments or instantiations of templates, functions, test
cases or altsteps and during the implicit assignment of the actual parameter to the formal
parameter. Examples of subtype violations are shown in Table 8.3. When such a subtype
violation occurs, it will be caught as a test case error. This is not necessarily a bad thing
and should not prevent you from using subtyping wherever applicable – indeed, the liberal
use of subtyping allows for early detection of illegal values, possibly even before the test
system is executed.

8.2.1   Type Aliasing

The simplest form of subtyping for a type is giving it a new name without restricting the
admissible set of values:

```
type integer HexadecimalInteger;
type integer OctalInteger;
```

```
type component SipUserAgent { port SipPort pt };
type SipUserAgent SipClient; // example alias of
                            // a user defined type
```

Type aliasing can be used for assigning intuitive names to a type that reflects its usage in a given context. It can also be used to allow a codec implementation to select suitable encodings based on, for example which alias of `integer` is to be encoded, `Hex-adecimalInteger` or `OctalInteger`. For example, a codec implementation could now differentiate between these types and encode the `HexadecimalInteger:255` as `"FF"`, whereas `OctalInteger:255` would be encoded as `"377"`.

8.2.2 Value List

A value list subtype restricts its values to a fixed list of allowed values that are explicitly enumerated in the type definition. A value list subtype can be defined for all types. Both literal expressions, like in the definition of `SipMethod`, and constants can be used, as shown in Table 8.4. It is also possible to mix constants and literals in the same type definition.

The use of constants in defining a value list subtype has the benefit that each allowed value of the type has an explicit name, which improves readability of the TTCN-3 code when these names are used consistently throughout the code.

8.2.3 Value Ranges

In many cases, in particular, for numerical types, enumerating the allowed values is not feasible, because there are simply too many of them. Already, when defining an 8-bit numerical type as a value list, the definition would be several lines long. Trying to use such a value list for the set of unsigned 64-bit integers is therefore not recommended! Instead, in such cases TTCN-3 allows the use of *value ranges* to define subtypes of float and integer types:

```
type integer SipStatusCode ( 100 .. 609 );
```

In the previous example, the boundary values are included in the range. To specify a value range where one or both boundary values are excluded, the corresponding boundary

Table 8.4 Defining value lists

```
type charstring SipMethod ( "REGISTER", "INVITE", "ACK", "BYE", "CANCEL",
                            "OPTIONS" );

const charstring c_REGISTER = "REGISTER";
const charstring c_INVITE   = "INVITE";

// further constants here ...
type charstring SipMethod2 ( c_REGISTER, c_INVITE, c_ACK, c_CANCEL, c_BYE,
                            c_OPTIONS );
```

has to be preceded by the character '!'. Therefore the previous example can also be written as:

```
type integer SipStatusCode ( 100 .. !610 );
```

Half-open ranges can be defined using infinity and -infinity. For example, since the content of a SIP message will never be of negative length, a better way to define SipContentLength would be:

```
type integer SipContentLength ( 0 .. infinity );
```

It is also possible to use constants in value range definitions, and to define value range subtypes for floats. Finally, it is also possible to combine value lists and ranges.

```
const float c_MaxTimeout := 20.0;
type  float Timeout ( 0.0 .. c_MaxTimeout );
type  integer SipInformationalStatusCode ( 100, 180 .. 183 );
```

8.2.4 Field Value Restriction for Structured Types

Earlier we introduced value lists as one form of subtyping that can also be used to further restrict values in structured type definitions. Quite often however these types have a large number of fields, for example records and sets, where the specification of value lists becomes quite cumbersome to specify and maintain when only the restriction of a small number of fields is of interest. To simplify such field value restriction TTCN-3 allows fields in such subtype definitions which are not of interest to be left out, see Table 8.5. The condition of such subtyping definitions are that field value restrictions do not conflict with value restrictions specified in the referenced parent type. Note that this form of subtyping can also be used with the anytype which will be introduced in our next chapter.

Table 8.5 Example field value restrictions for structured types

```
type record SipStatus {
  float           version,
  integer         statusCode,
  charstring      reasonPhrase
};

type record SipStatus SipStatus100Trying {
  integer   statusCode (100),
  charstring reasonPhrase ("Trying")
};

type record SipStatus SipInformationalStatus {
  integer   statusCode (100, 180)
};

type record SipStatus100Trying SipInformationalStatusError {
  integer   statusCode (100, 180) // 180 violates parent type restriction!
};
```

8.2.5 Type Lists

The type list notation provides a convenient way to specify a new subtype which is based on two or more existing subtypes. The type list notation can be used for basic types, structured types and anytype. The types in the list shall be subtypes of the same root type. The new subtype defined by this list restricts the allowed values of the subtype to the union of the values of the listed subtypes. An example of the type list notation is shown in Table 8.6. The new subtype `SipMethod1and2` allows the union of all the values in the associated value lists of the subtypes `SipMethod` and `SipMethod2`.

Type lists may also contain additional length restrictions for list types or value list restrictions for any structured type or the `anytype` as shown in our example in Table 8.6. In the case of `length` restrictions, the keyword has to follow the subtype identifier. Similar to the case of field value restrictions, these additional value constraints must not violate any of the value constraints set in any of its parent types.

8.2.6 Character Set Restrictions for Strings

It is possible to restrict the set of allowed characters for a string value using a character set restriction. For example, the first subtype in Table 8.7 shows one possible way to restrict a charstring type to only alphanumerical characters.

It is also possible to list single characters or combine single characters with character ranges. A definition for the string type that can represent SIP URIs (Uniform Resource Identifiers) is given in Table 8.7. URIs are used to identify communication partners. The strings `"sip:alice@atlanta.com"` and `"sip:bob@biloxi.com"` from the SIP example in Table 8.1 are valid textually encoded SIP URIs.

As you can see, character constants may be used in the type definitions interchangeably with character literals, similar to the numerical ranges.

8.2.7 Length Restrictions for Strings and List Types

Many protocols, in particular, those whose PDUs are designed to be carried by packet-oriented transport protocols, often place upper boundaries on the length of strings and other list data types. TTCN-3 allows these restrictions with the `length` keyword to be expressed. It is possible to specify the exact length or a length range; `infinity` can be

Table 8.6 Defining of and subtyping with type lists

```
type charstring SipMethod ("REGISTER", "INVITE", "ACK", "BYE", "CANCEL",
                           "OPTIONS" );
type charstring SipMethod2 (c_REGISTER, c_INVITE, c_ACK, c_CANCEL, c_BYE,
                           c_OPTIONS );
type charstring SipMethod1and2 (SipMethod, SipMethod2 );
type record length(1..5) of integer OneToFiveIntegers;
type OneToFiveIntegers ThreeIntegers length (3); // type list subtype OK
type OneToFiveIntegers TenIntegers length (10); // ERROR - violation of
                                                // length constraint
```

Table 8.7 Character set restrictions

```
type charstring AlphaNumString ( "0" .. "9", "A" .. "Z", "a" .. "z" );

const charstring c_at  := "@";
const charstring c_dot := ".";
type charstring EncodedSipUri ( "0" .. "9", "A" .. "Z", "a" .. "z", " ",
                                "[", "]", ":","&", "=", "+", "$", ",",
                                ";", "?", "-", "_", "/", "%", c_at,
                                c_dot, "'", "'", " " );
```

used to specify upwardly open ranges. For length restrictions it is not possible to exclude boundary values as shown for value ranges in Section 8.2.3.

```
// exactly 1024 characters
type charstring Payload          length ( 1024 );
// 0 to 1024 characters
type charstring VariablePayload length ( 0 .. 1024 );
// 1 or more characters
type charstring NonEmptyString  length ( 1 .. infinity );
```

Length restriction subtypes can also be specified for the other TTCN-3 string types (universal charstring, bitstring, hexstring and octetstring). It is important to note that the length unit varies from string type to string type. See Table 8.8 for details and examples.

Finally, it is also possible to define length-restricted subtypes for the list types record of and set of. Examples for this are shown below. Note that the syntax is slightly different in that the length restriction is not specified at the end of the type declaration but directly after the record or set keyword.

```
type record length ( 4 ) of integer Quadruple;
type set length ( 1 .. infinity ) of SipMethod SipMethods;
```

Table 8.8 String length units

String type	Length unit	Example
charstring	Character	"Bob <sip:bob@biloxi.com>" is 24 characters long
universal charstring	Universal character	"200?" is 4 universal characters long
bitstring	Bit	'101101'B is 6 bits long
hexstring	Hexadecimal digit	'FFEF7B'H is 6 hexadecimal digits long
octetstring	Octet	'FFEF7B'O is 3 octets long

8.2.8 Subtyping of Subtypes

In the above examples, we have derived the subtypes directly from built-in types such as `integer` and `charstring`. It is also possible to derive subtypes from existing defined subtypes. For example, if we wished to reflect the fact that an informational SIP response status code is a specific kind of status code, we could define the following.

```
type SipStatusCode SipInfoStatusCode ( 100, 180 .. 183 );
```

The restrictions imposed on the newly derived subtype and the parent type are cumulative, that is a value of type `SipInfoStatusCode` must be an `integer` between 100 and 609 that either equals 100 or lies between 180 and 183. In this example, no additional constraint is imposed on the values of `SipInfoStatusCode` by deriving it from `SipStatusCode`: the values allowed by the new restriction are all admissible values of `SipStatusCode`. Of course, this is not necessarily the case. It should, however, be considered good style because it certainly improves the readability of the code.

Since refering to undefined types is not allowed, the parent type of a subtype must be defined and there must be no cycles in a chain of subtype definitions:

```
type SubType1 SubType2;
type SubType2 SubType1; // ERROR: cycle in subtype definition
```

Note that this restriction refers to all cycles, not just cycles with length 2.

8.2.9 Type Conversion

TTCN-3 does not allow for any form of implicit type conversion or 'casting' as in some other programming languages. Not even mixing `integer` and `float` in arithmetic expressions is allowed. For example, `10 + 2.0` is rejected as ill-typed. Instead of implicit type conversion, TTCN-3 offers a rich set of explicit conversion functions that allow for controlled conversion between various types. Most of these pre-defined functions will be introduced in the following sections. For a complete, detailed list of these functions, refer to the language standard [8, Annex C].

8.3 TTCN-3 Built-in Types

TTCN-3 is used in the testing of a wide range of different systems. To be able to represent the data and messages of these diverse systems, TTCN-3 possesses a large number of built-in types and ways to define new types from the existing ones. We will now introduce the various TTCN-3 built-in types together with their operations and some useful functions. Every type, either built-in or user defined, can be used in (in-)equality tests using the operators '==' or '!=', respectively. Comparing values of equal type or at least stemming from the same type via a sequence of subtype definitions is allowed, as is shown in Table 8.9. If it is necessary to compare different types, the TTCN-3 conversion functions [8, Annex C] should be used to make explicit conversions.

Table 8.9 Comparing type-incompatible values is a type error

```
var integer          v_i  := 0;
var SipContentLength v_cl := 132;
var float            v_f  := 0.0;

// TYPE ERROR
if  ( v_i != v_f ) { /* ... */ }

// OK: i and cl stem from integer
for ( v_i := 0 ; v_i < v_cl; v_i := v_i + 1 )
```

TTCN-3 provides a large set of basic types that are well known from other classical programming languages. There are a number of differences, though, which will be highlighted as we explain the types. TTCN-3 also has a number of additional types that distinguish it as a test scripting language with a strong focus on protocol testing. We will describe these types once we have covered the more conventional types.

8.3.1 The Boolean Type

TTCN-3 has a genuine `boolean` built-in type, which can assume the two truth values `true` and `false`. The Boolean operators `and, or, xor` and `not` can be used to form Boolean expressions. In addition, it should be noted that each equality or comparison operation also evaluates to a Boolean value, see Table 8.10 for example.

8.3.2 The Integer Type

TTCN-3 provides only a single built-in type for integral numbers: `integer`. Values of type `integer` can be arbitrarily large. This is different from other programming languages, where there are usually a number of integer types with different admissible value ranges. In general, however, it is not recommended to use integer to store arbitrary large, non-numerical data as most TTCN-3 tools support only signed 32-bit or signed 64-bit integer values rather than arbitrarily large integers. Such data is better represented using one of the binary string types (Section 8.3.6), which also have a number of useful

Table 8.10 Example of Boolean variables and Boolean operators

```
const boolean c_discardMessages      := true;
const boolean c_earlyDiscardMessages := false;
type integer MaxForwards ( 0..infinity );
var MaxForwards v_maxForwards := f_getMaxForwardsHeaderValue( ... );

var boolean v_dropMessage      := ( ( c_discardMessage and
                                    ( maxForwards <= 0 ) ) or
                                    ( c_earlyDiscardMessage and
                                    ( v_maxForwards <= 10 ) ) );
```

Table 8.11 Useful integer subtypes

```
type integer byte              ( -128 .. 127 );
type integer unsignedbyte      ( 0 .. 255 );
type integer short             ( -32768 .. 32767 );
type integer unsignedshort     ( 0 .. 65535 );
type integer long              ( -2147483648 .. 2147483647 );
type integer unsignedlong      ( 0 .. 4294967295 );
type integer longlong          ( -9223372036854775808 ..
                                   9223372036854775807 );
type integer unsignedlonglong ( 0 .. 18446744073709551615 );
```

operations for the processing of non-numerical, binary data that are not available for
`integer`. When `integer` values are used to store data to be sent or received from
the SUT, it is good practice to use suitable subtype restrictions that reflect the supported
value ranges of the SUT. Examples of such are given in Table 8.11, a list of additional
useful Types, based on other built-in types of TTCN-3, can be found in [8, Annex E].

`integer` is an ordered type, so, in addition to (in-)equality, integers can be compared
with the operators '<', '>', '<=' and '>=' with their usual interpretation. The arithmetic
operators '+', '-', '*' and '/' are available, where '/' denotes integral division, so that, for
example 7 / 2 yields 3.

The remainder of integral division is given by the operator 'rem', so that 7 rem 2
yields 1. In general, for two given non-zero values n and m, n rem m yields n - (n /
m) * m. There is also a modulo operator mod that differs from rem in that n mod m is
the smallest *positive* number such that m divides n - (n mod m). Table 8.12 shows a
TTCN-3 implementation of (a slight optimisation of) Euclid's algorithm, which calculates
the greatest common divisor of two positive `integer` values.

Division by zero is a test case error and will cause termination of the currently executing
test case. For example, calling f_gcd(17,0) will cause a test case error because the
calculation 17 rem 0 involves a division by zero.

Table 8.12 A slightly improved version of Euclid's algorithm for the greatest common divisor

```
type integer PositiveNumber ( 0 .. infinity );

function f_gcd( in PositiveNumber p_n, in PositiveNumber p_m )
return PositiveNumber {
  while ( p_n != p_m ) {
    if ( p_n < p_m ) {
      p_m := p_m rem p_n;
    }
    else {
      p_n := p_n rem p_m;
    }
  }
  return p_n;
}
```

For the conversion to `integer` values TTCN-3 offers a number of pre-defined functions which include `float2int`, `char2int`, `unichar2int`, `bit2int`, `hex2int`, `oct2int` and `str2int`. The difference between `char2int` and `str2int` functions is that the first converts a single character into its decimal representation, that is it returns `80` for `"P"`, whereas the latter maps numeric characters into an `integer` value, that is it returns `-123` for `"-123"`. The passing of incorrect arguments to any of these functions, for example a string containing letters to `str2int`, leads to a test case error.

8.3.3 The Float Type

Real numbers in TTCN-3 are represented by values of type `float`. The TTCN-3 standard does not make a statement about required precision for `float` values, but since arbitrary precision real numbers are virtually impossible to achieve, your TTCN-3 tool will probably place a practical limit on available precision. Here are some examples of float literals:

```
const float c_sipVersion = 2.0;
const float c_pi         = 0.03141592E2;
            // pi = 0.03141592... * 10^2
```

As you can see from the examples, scientific notation to base 10 is also supported.

Like `integer`, `float` is an ordered numerical type that has comparison and arithmetic operators. It is worth stressing again that both the arithmetic operations and comparisons must not mix values of `integer` and `float` without explicit conversion by means of the pre-defined conversion functions `int2float` and `float2int`.

The function `float2int` rounds its argument `float` by stripping the fractional part. Rounding to the nearest `integer` is shown in Table 8.13. In addition, `float` values can be produced from character strings complying to obvious constraints by using the `str2float` conversion function.

It should be noted that care is needed when using `float` expressions – rounding errors may lead to unexpected results. For example, the expression `(10.0 / 3.0) * 3.0 == 10.0` may evaluate to `false` depending on the error introduced when the intermediate result `3.33333...` is stored in a finite-precision representation. Similarly, using `float` templates to match incoming messages may lead to unexpected results. With this in mind, you see that the choice to model the SIP version as a `float` value is actually a very poor choice, and we will see a more appropriate modelling in a later example.

Table 8.13 Using float2int

```
function f_round( in float p_x ) return integer {
  if ( p_x >= 0.0 ) {
    return ( float2int( p_x + 0.5 ));
  }
  else {
    return ( float2int( p_x - 0.5 ));
  }
}
```

TTCN-3 does not provide any mathematical functions like (co)sine, exponential or logarithm. However, one useful function that the language does provide is random number generation. The function `rnd` generates pseudo-random numbers between `0.0` and `1.0`. The function has an optional `float` seed value. When called without seed value, the return value of the previous invocation of `rnd` will be used for seeding. This means that pseudo random number sequences seeded with the same initial value will be identical. This can be useful when attempting to write test cases that perform randomised tests in a reproducible manner. If the first invocation of `rnd` is without a seed value, then the sequence of random numbers will not be predictable and therefore not be reproducible. The following example shows how pseudo-random `integer` values from a given range can be calculated on the basis of the pseudo-random numbers generated with `rnd`.

```
function f_rnd_integer( in integer p_lower, in integer p_upper )
return integer {
  return float2int( rnd() * int2float( p_upper - p_lower + 1 ) )
        + p_lower;
}
```

8.3.4 The Charstring and the Universal Charstring Type

TTCN-3 has two different character string types `charstring` and `universal charstring`. The `charstring` type is restricted to represent 7-bit ASCII strings and as such can be used to model the most common human readable text in Latin alphabet. The `universal charstring` type is far more extensive and may contain characters from the Unicode character set described in ISO/IEC 10646 [33].

String literals are enclosed in double quotes (`"`), as shown in Table 8.14. TTCN-3 does not support escape sequences (\n, \t, \", ...) for non-printable characters. The only exception is the double quote (`"`), which can be inserted into strings using a pair of double quotes (`""`). Thus, `charstring` literals can only be a sequence of the *printable* 7-bit ASCII characters (from SP (32) to TILDE (126)). The conversion function `int2char` can be used to generate non-printable characters from their `integer` encoding. A `universal charstring` can contain virtually every printable symbol that exists, although it will depend on your TTCN-3 tool and/or text editor if you can use them in your program code directly, or if you will have to use the quadruple notation, which specifies a `universal charstring` character by its group, plane, row and cell specified in ISO/IEC 10646 [33].

Individual characters of a string can be read and (in case of string variables) set with the subscription operator `"[]"`, with indices starting from 0. Attempts to read beyond

Table 8.14 Some examples for charstring and universal charstring literals

```
var charstring v_callerName               := "alice";
var charstring v_callerRealname           := "Alice ""Host"" Bell";
var universal charstring v_finnishCallerName := "Yrjö Åberg"
var universal charstring v_euroSign        := char(0, 0, 32, 170); // = €
```

the boundaries of a string will cause a test case error, on the contrary it is possible to append a character to a string by assigning a character at the corresponding index position. Examples of valid character accesses shown here:

```
if ( v_callerName[0]      == "a" ) ...        // evaluates to true
if ( v_callerRealName[6]  == """" ) ...       // evaluates to true
    v_callerName[4]       := a; // evaluates to "alica"
    // extending a string by one character, evaluates to "alicax"
    v_callerName[5]       := "x";
    // evaluates to "Yrjö Åberg"
    v_finnishCallerName[5] := char( 0,0,0,216 );
```

What can also be seen from these examples is that single characters, both of charstring and universal charstring, are also written between double quotes – there is no special form of quoting for character types. Note that some earlier versions of the TTCN-3 language contained separate character types, these types are now considered as synonyms for (universal) charstrings of length 1 and are included in the language standard as part of generally useful type definitions in [8, Annex E].

String concatenation is performed with the concatenation operator "&". The length of a string is returned by the lengthof operation. Table 8.15 shows how concatenation and the lengthof operation can be used to append at the proper place the line termination (CR/LF) to a charstring value. The lengthof operation can be used with constant and variable values. The passing of incorrect arguments to this function, for example templates that accept strings of variable length, leads to a test case error.

Among the various pre-defined type conversion functions of TTCN-3, there is no direct function that allows the conversion between charstring and universal charstring. Similarly, there is no function to convert float values to string values. Fortunately, these conversions can all be easily accomplished by first converting to an integer value as an intermediate step. Next to the int2char function TTCN-3 offers also int2unichar and oct2char functions to convert values of other types to

Table 8.15 Appending CR/LF to a charstring

```
const charstring c_CR := int2char( 13 );
const charstring c_LF := int2char( 10 );

function f_appendCRLF( in charstring p_str ) return charstring {
  return ( p_str & c_CR & c_LF );
};

function f_appendCRLF2( in charstring p_str ) return charstring {
  var integer len       = lengthof( p_str );
  var charstring result = p_str & "XX"; // make room for CR/LF
  result[len]           = c_CR;
  result[len+1]         = c_LF;
  return result;
}
```

`charstring` and `universal charstring` values. The use of incorrect arguments with these functions, for example a negative integer or an integer greater than 127 with `int2char`, leads to a test case error.

In addition to the ability to access single string elements with the subscription operator, TTCN-3 provides the pre-defined function `substr`, which can be used to extract a substring from a given string value or template. In addition to the string itself, `substr` takes two more arguments: the start index of the substring (with indices starting from zero), and the length of the substring. So, for example `substr("federico.engler",0,8)` yields `"federico"`. The return type of `substr` is the type of its string argument. The passing of incorrect arguments to this function, for example a start index outside the string's admissible range or a negative length value, leads to a test case error.

The predefined function `replace` can be used to replace or insert a specified substring in a string value or template. Next to the string which is to be modified, `replace` takes three more arguments: the start index of the substring to be replaced (with indices starting from zero), the length of the substring to be replaced and the string to be inserted. A length value of zero results in an insertion. So, for example `replace("engler",0,0,"federico.")` yields a return value of `"federico.`
`engler"`. The return type of `replace` is the type of its string arguments. The passing of incorrect arguments to this function, for example a negative index value or use of different string types for the two string arguments, leads to a test case error. The pre-defined functions `lengthof`, `substring` and `replace` are not only applicable to `charstring` but also to `bitstring`, `hexstring` and `octetstring`, the list types `record of` and `set of` and arrays.

8.3.5   The Verdicttype Type

TTCN-3 has a type to represent the possible outcomes – verdicts – of a test case, which is called `verdicttype`. It has five possible values: `none`, `pass`, `inconc`, `fail` and `error`. We have already mentioned that each test component implicitly carries a value of type `verdicttype`, which stores the current local verdict. This state can be set with `setverdict` and can be read with `getverdict`.

```
setverdict( pass );
:
if ( getverdict() == fail ) { /* ... */ }
```

These operations and the related verdict overwriting rules have already been discussed in Section 4.3.2 and will not be repeated here. The type `verdicttype` does not have any operation except the check for (in)equality.

8.3.6   The Binary String Types Bitstring, Hexstring and Octetstring

Raw, binary data is best represented in TTCN-3 using its different binary string types. Depending on the desired data alignment, the types `bitstring`, `hexstring` or

octetstring can be used. These string types allow the representation of binary data
either without grouping, with grouping of 4 bits or with grouping of 8 bits, respectively.
Most functions and operators for these types are similar (and indeed already known from
the character string types), so we will introduce them here only briefly, highlighting
peculiarities and differences where they exist.

Literals for the binary string types are written as a (possibly empty) sequence of binary
(for bitstring) or hexadecimal digits (hexstring, octetstring) in single quotes
(') followed by the letters 'B', 'H' or 'O'.

```
const bitstring   c_2005_bit := '11111010101'B;
const hexstring   c_2005_hex := '7d5'H;
const octetstring c_2005_oct := '07D5'O;
```

As can be seen from the examples, the case of the hexadecimal digits 'a'-'f' does not
matter and can even be mixed freely in a single literal. Octet string literals must consist
of zero or an even number of hexadecimal digits – two hexadecimal digits correspond to
one octet (8 bits).

Like for the character strings, access to individual string elements is provided with the
subscription operator ([ ]). Indices start from zero, but represent different units depending
on the type of the binary string: the subscription index represent binary digits, hexadecimal
digits or pairs of hexadecimal digits, respectively. Attempts to access digits past the right
end of a string leads to a test case error.

```
c_2005_bit[1]              // evaluates to '1'B
c_2005_hex[1]              // evaluates to 'D'H
c_2005_oct[1]              // evaluates to 'd5'O
lengthof( '11111010101'B ) // evaluates to 11
lengthof( '7D5'H )         // evaluates to 3
lengthof( '07d5'O )        // evaluates to 2
```

To guard against this error, the length of a binary string can be retrieved by the
lengthof function, which returns the length of a string measured in binary, hexadecimal or pairs of hexadecimal digits, depending on the string type. For the conversion of
values to binary string values TTCN-3 offers a number of pre-defined functions which
include int2bit, int2hex, int2oct, char2oct, bit2hex, bit2oct, hex2oct,
oct2bit, oct2hex and str2oct. The passing of incorrect arguments to any of these
functions, for example a string with an odd number of characters to str2oct, leads to
a test case error.

8.3.6.1 Operations for Binary String Types

TTCN-3 has a number of operators that perform bitwise Boolean operations on values of
binary string types: not4b, and4b, or4b and xor4b, which read as 'not for bit', 'and
for bit' and so on. The operator not4b takes one argument of arbitrary size, the other
operators require two arguments of the same binary string type and same length. The result
is again of the same length and binary string type – it is obtained by combining the binary

representation of the argument(s) bit by bit with the respective Boolean operations – a more comprehensive example is given in Section 15.4.6.

```
not4b '101101'B        // evaluates to '010010'B
not4b 'EF'H            // evaluates to '10'H
'101101'B and4b '111000'B // evaluates to '101000'B
'AC'O and4b 'FF'O      // evaluates to 'AC'O
```

The left shift (<<) and the right shift (>>) operators shift binary string values by a positive number of units in the specified direction. Again, the unit depends on the string type, it is a single bit for `bitstring`, a hexadecimal digit for `hexstring` and two hexadecimal digits for `octetstring`. Freed positions in the string are filled with 0 and excess units are shifted out of the string, so shifting does not alter the string's length. The left rotate (<@) and right rotate (@>) operator, work in a similar way, but fill the freed positions with the excess digits that are shifted out of the string. Both for shift and rotate, if the second operand is larger than the length of the string, a test case error will occur.

```
'11001011'B << 3 // evaluates to '01011000'B
'11001011'B <@ 3 // evaluates to '01011110'B
'abcd01'H   << 2 // evaluates to 'cd0100'H
'abcd01'H   <@ 3 // evaluates to 'cd01ab'H
'FF01EF'O   << 2 // evaluates to 'EF0000'O
'FF01EF'O   <@ 2 // evaluates to 'EFFF01'O
```

8.3.7 The Default Type

TTCN-3 allows activating and deactivating `altsteps` as default behaviour, see Section 4.9. To be able to deactivate specific default altsteps, default references are returned by the associated `activate` statement. These act as a handle to the newly activated default and can then be used to deactivate them. These default references are treated as normal values of type `default`. They can be assigned to variables and then be passed in and out of functions. The literal `null` can be used to initialise values of type `default`. The only other way to obtain a `default` value is by calling `activate`. Section 15.4.3 shows how `default` variables can be used to achieve guaranteed uninterrupted sleep of a component even in the presence of possibly active 'catch all' defaults.

8.4 User-Defined Types

In addition to the built-in types of TTCN-3, there exists a number of ways to define new types from existing ones. Some of these ways are well known from other programming languages, such as enumerations, records and unions. Others are less well known, but prove to be equally useful in many test situations. In the following sections, we will cover the different ways to create user-defined types in TTCN-3, and give examples of what kind of data structures from test applications or protocols can be modelled with these types.

8.4.1 The Enumerated Type

Enumerated types are an ideal way to represent types that have a small, finite set of values. For example, encoded DNS messages indicate their message kind by having a single bit set either to zero, in case of a question, or to one in case of an answer. In our DNS message definition, it is a good idea to represent these different message kinds not by a bitstring or integer representation but instead by assigning abstract, *symbolic* names to the different DNS message kinds, as shown in Table 8.16. It is not allowed to have the same element occur more than once in a single enumerated type. It is possible though to have the same element name occur in different enumerated types.

It is possible to assign explicit numbering to the enumeration elements as shown in the example EuroCoins in Table 8.16. Such numbering can then be used as a base for comparisons between enumeration elements. Enumerations are ordered types and hence can be compared using the comparison operators "<", ">", "<=" and ">=". If no explicit numbers are assigned to an enumeration's element, the elements are numbered in ascending order from 0 in their order of occurrence in the type definition. It is also possible to assign explicit numbers to only some of an enumeration's elements. In this case, the remaining elements are numbered increasingly from 0 using those numbers that are not explicitly assigned. This, however, will lead to a hard to grasp arrangement of the elements and hence should be avoided.

Note that although the numerical values for the elements are used typically only to define their order there is a predefined function in TTCN-3 to return the numerical value corresponding to an element of an enumerated type. The predefined functions enum2int and int2enum take the enumerated value as its argument and return the associated integer value, and vice versa.

8.4.2 The Record Type

Records can be used to group related fields into a single type. For example, host addresses and SIP URIs could be stored in a structured way using the record definitions from

Table 8.16 Defining and using enumerated types

```
type enumerated DnsMessageKind {
  e_question, e_answer
    // will be interpreted as: e_question( 0 ), e_answer( 1 )
}

type enumerated EuroCoins {
  e_1cent( 1 ), e_2cent( 2 ), e_5cent( 5 ), e_10cent( 10 ), e_20cent( 20 ),
  e_50cent( 50 ), e_1euro( 100 ), e_2euro( 200 )
}

var EuroCoins v_coin := f_getCoin( ... );
if ( v_coin < e_10cent ) {
  log( "Coin made of copper!" );
}
```

Table 8.17   Record type definitions and values

```
type record SipStatus {
  float           version,
  SipStatusCode  statusCode,
  charstring      reasonPhrase
};

const SipStatus c_successStatus := {2.0, 200, "OK"};   // value list

const SipStatus c_failureStatus := {                    // assignment list
  version      := 2.0,
  statusCode   := 400,
  reasonPhrase := "Bad Request"
};
```

Table 8.17. Field names within a record must be unique but may be re-used in different record type definitions. Each field may either be of a built-in or user-defined type. A field type definition must not contain a reference to another field type. Table 8.17 also shows the different notations available to specify record values: value list notation and assignment list notation. The value list notation specifies values for all the fields of the record in their order of occurrence in the type definition, whereas the assignment list notation explicitly specifies the field names. As you can see, the assignment list notation makes for a more structured but also more verbose value definition. For records with more than, say, four fields, the assignment list notation is strongly advisable to keep track of record's fields. Also, in general, assignment list notation will improve the readability of your code. Fields in the assignment list notation do not have to appear in the order that has been used in the type definition, but using this convention will improve the readability of your code.

Individual fields of a record value can be accessed for reading and writing using the dot operator " . " together with the field names, as shown in the following example. For nested record types, that is records that have fields of record type, inner fields can be accessed by chaining the field names.

```
var SipStatus failureStatus := c_failureStatus;
failureStatus.statusCode    := 404;
failureStatus.reasonPhrase  := "Not Found";

if ( c_successStatus.statusCode == 200 ) { ... }
```

It is possible to specify only partially defined record values. In the assignment list notation, this is achieved by simply leaving out the fields that are to be left undefined. For the value list notation, the symbol " - " must be used for the undefined fields. Partially defined values are only allowed in a transient state, that is only as long as the undefined fields are not used in some operation. Reading an undefined field will cause a test case error; the same holds for using a partially defined value in a comparison operation. It is also not permissible to use a partially defined value in a send-like statement. On the other hand, it is possible to use partially defined values in assignments and to pass them

Table 8.18   Partially defined record values

```
var SipStatus partialStatus  := {2.0,-,-};

var SipStatus partialStatus1 := {
  version := 2.0
};

partialStatus.statusCode       := 100;                          // OK - write
var charstring v_reasonPhrase := partialStatus.reasonPhrase; // ERROR - read
if ( partialStatus == partialStatus1 ) { ... }                  // ERROR - read
```

in and out of functions as reference parameters (out or inout parameters). Examples
of partially defined record values are shown in Table 8.18.

8.4.2.1   Optional Fields

We have already touched on the subject of having partially defined values and the problems caused by them. In most occasions, it is more appropriate to use optional fields to
model records where some fields may be missing in certain situations. SIP URIs of the
form "sip:bob:myPasswd@biloxi.com:8081", where username, password or port
number may be left unspecified in certain contexts could be modelled as a record type as
shown in Table 8.16. There are two alternatives to indicate that an optional value shall
be absent. Firstly, instead of a value the keyword omit can be used. In Table 8.19 the
URIs sip:alice@atlanta.com or sip:atlanta.com:8081 are written as values
of type SipUri. Note that the type definitions for these examples do not show the additional encoding information, for example the "@" following the user information in a SIP
URI, required to arrive at the final encoded string values. For more information about our
approach taken for textual encoding, refer to Section 7.5.4.

Secondly, it is possible to declare that in the definition of a value of a record type, all
optional fields that are not explicitly assigned a value are implicitly set to omit. This will
result in shorter definitions, especially if there are many optional fields without a value.
On the other hand, forgetting to set an optional to either a specific value or to be absent
can no longer be caught by TTCN-3 analysers. This way of implicitly setting optional
fields to omit is available for constants, variables, templates and module parameters and
it can be applied from single definitions to all definitions in a module. By default, TTCN-3
expects that optional fields are explicitly set to omit, to change this to implicit omit the
attribute mechanism has to be used, see Section 7.5.

Record values with omitted fields are fully defined; the fields can be freely read, the
values can be used in comparisons, and values with omitted fields can be freely used
in send-like statements. The only difference between omit and a regular value is that
omit is only valid in the context of an optional record field. Assigning an omitted field
to a regular variable or non-optional record field will cause a test case error. The pre-
defined function ispresent can be used to guard against these situations and to check
whether an optional field is present in a record value, as shown in the following examples.

Table 8.19 Record types with optional fields

```
type record HostPort {
  charstring    host,                    // user domain
  unsignedshort portNumber optional      // optional port
};

type record UserInfo {
  charstring name,
  charstring password optional
};

type record SipUri {
  UserInfo    userInfo optional,    // optional user name and password
  HostPort    hostPort,             // user domain and port
  charstring uriParams optional,    // an encoded list of URI parameters
  charstring headers optional       // an encoded list of URI headers
};

const SipUri c_uriAlice := {
  userInfo  := { "alice", omit },       // omit in value list notation
  hostPort  := { "atlanta.com", omit },
  uriParams := omit,
  headers   := omit
};

const SipUri c_uriAtlanta := {
  userInfo  := omit,                     // omit in assignment list notation
  hostPort  := {"atlanta.com", 8081},
  uriParams := omit,
  headers   := omit
};
```

In addition, the pre-defined function `sizeof` can be used to determine the number of present fields in a `record` value or template.

It should be noted that in addition to constant and variable values `ispresent` can also be used to check fields of template instances of record type which are introduced in the next chapter. Users have to be aware that the incorrect use of this function, for example use with an inaccessible field, leads to a test case error.

```
c_uriAlice.userInfo.password == "mySecretPassword" // evaluates to
                                                    // false
var UserInfo v_userInfo := c_uriAtlanta.userInfo;   // ERROR
template UserInfo a_anyUserAndPassword := { name := ?,
                                            password := ? };

ispresent( c_uriAlice.userInfo )     // evaluates to true
ispresent( c_uriAtlanta.userInfo )   // evaluates to false
sizeof( c_uriAtlanta )               // returns 1
sizeof( a_anyUserAndPassword )       // returns 2
```

8.4.3   The Set Type

Sometimes, the strict order of fields in a record is not necessary and not in line with the actual usage of the data in the tested application or protocol. One example is the digest data being exchanged between SIP entities to achieve authentication of the communication partners. This exchange is based on a challenge-response mechanism, in which one party proves its identity on the basis of a secret (usually a password), that is shared between the communication partners. To avoid sending this secret in clear-text over the network, a challenge based on a fresh nonce value is sent to the party that shall authenticate itself. A valid response to this challenge is a hash value of this nonce and the shared secret. This can be checked by the challenging party and thus proves the identity. A cryptographically secure hash function should be used to make guessing of the shared secret from the hash value unfeasible.

SIP messages do not require a specific order of the values sent in a digest challenge, so it would be un-natural to require such an order in the corresponding TTCN-3 type. Table 8.20 shows how a (slightly simplified) digest challenge could be represented as a set type in TTCN-3. In this example, optional fields are implicitly omitted. It is assumed that for the complete module implicit omit has been set via the use of the optional attribute, see Section 7.5.

On the level of TTCN-3 code, there is only a minimal difference between sets and records, therefore nearly all that we have described for records in Section 8.4.2 can equally be applied to sets. The only major difference is that set values may not be written using the value list notation.

The main conceptual difference between sets and records is that records should be used to represent structured values whose fields have to be encoded in a fixed order, whereas sets should be used where the fields may be encoded in arbitrary order.

8.4.4   The Union Type

Records and sets combine groups of different types where all the grouped types are always present. In some situations, it is also useful to have a type that combines a group

Table 8.20   Defining and using set types

```
type charstring DigestAlgorithm ( "MD5", "MD5-sess" /* ... */ );

type set DigestChallenge {
  charstring       realm,           // what we authenticate ourselves for
  charstring       nonce,           // the fresh value for this challenge
  DigestAlgorithm algorithm,        // the algorithm to compute the hash value
  charstring       opaque optional  // opaque data to be returned with
                                    // response
}

const DigestChallenge c_md5Challenge := {
  realm      := "atlanta.com",
  nonce      := "QW50amVKb2hhbm5hQW5uaWthTWlhU3RlcGhhbg==",
  algorithm := "MD5",
  // the field 'opaque' is not mentioned and implicitly set to omit
}
```

Table 8.21 Defining and using union types

```
type union SipRequestUri {
  SipUri    sip,
  SipsUri   sips,
  TelUri    tel,
  FaxUri    fax,
  ModemUri modem
};

type charstring TelUri;
type charstring FaxUri;
type charstring ModemUri;

const SipRequestUri c_NYTelUri := {
  tel := "+212.111.4444"
};

const SipRequestUri c_BobSipUri := {
  sip := {
    userInfo  := { "bob", omit },
    hostPort  := { "biloxi.com", omit },
    uriParams := omit,
    headers   := omit
  }
};
```

of different types in a way that exactly one of these types is present at any one time. In TTCN-3, this is achieved with union types. For example, the URI, which is used in the first line of a SIP request, may be a SIP URI but can also be of another type such as telephone, modem or fax URIs. A type, that is capable of representing one value of the different URIs in a single place could be defined as a union of these URI types. An example of this is shown in Table 8.21. Values of type SipRequestUri can now contain one of the five possible constituent types of the union. We will refer to these different possibilities as the *variants* of the union type. Values for union type are written in assignment list notation with only a single field, which specifies the variant.

Access to a variant of a union value is done with the dot operator '.' like field access for record values. It is important to note that TTCN-3 unions are so-called *tagged* unions: once a variant has been selected, this is recorded in the value and attempts to read a different variant will result in a test case error. This is even the case if the set variant and the variant, that is attempted to be read are of the same type. Also it should be noted that values of the same union type with different selected variants are never equal. However, it is possible to change the selected variant of a union value in an assignment, which will override the previously selected variant and make the previously stored value inaccessible.

```
var SipRequestUri v_telRequestUri := { tel := "+212.111.4444" };
var SipRequestUri v_faxRequestUri := { fax := "+212.111.4444" };
```

```
if ( v_telRequestUri == v_faxRequestUri ) ...
  // evaluates to false since tel and fax
  // variants are compared

// OK - tel is selected variant
var charstring x := v_telRequestUri.tel;
// ERROR - fax is not selected!
var charstring y := v_telRequestUri.fax;

// OK - overrides set variant
v_telRequestUri.fax := "+212.222.2222";
// OK now
var charstring z := v_telRequestUri.fax;
// ERROR now
var charstring n := v_telRequestUri.tel;
```

The pre-defined function ischosen can be used to determine if a union value or template has a certain variant selected. This can be used to prevent illegal access of the form shown above. An example for the use of ischosen to examine the selected variant of a union is given in Table 8.22.

As in the case of ispresent, the ischosen function can also be used with constant values and templates. Its incorrect use, for example, with an inaccessible field, also leads to a test case error.

8.4.5 The List Types

So far, we have studied ways to group collections of different types into structured types. TTCN-3 also supports different ways to collect a bounded or unbounded number of values of the same type into one value: arrays and record-of types for ordered groups, and set-of types for unordered groups. What all these list types have in common is that a number of operators and pre-defined functions defined in earlier sections in the context of character and binary strings can also be used with list values. These operators include rotation '<@' and '@>', and concatenation '&'. Pre-defined functions that can be also used with list values and templates include lengthof, substr and replace functions.

Table 8.22 Using ischosen

```
function f_isSipRfc3261RequestUri( in SipRequestUri p_requestUri )
return boolean {
  if ( ischosen( p_requestUri.sip ) or ischosen( p_requestUri.sips ) ) {
    return true;
  }
  else if ( ischosen( p_requestUri.tel ) or ischosen( p_requestUri.fax ) or
            ischosen( p_requestUri.modem ) ) {
    return false;  // since defined in RFC 2806
  }
}
```

8.4.5.1   The Record-of Type

The most natural way to define an ordered collection of elements – lists or vectors – of the same type in TTCN-3 is the record-of type. Record-of types may contain an arbitrary number of elements, but may be subtyped to fixed length or length ranges. The length ranges always contain the boundary values.

For example, SIP allows requests and responses to carry an arbitrary number of headers in a fixed order. Therefore, an appropriate way of modelling such headers is with a `record of` type. It is also possible to represent an IP address using such a type as shown in Table 8.23. Record-of values are shown in our examples using the value list notation. It is possible to use the symbol `"-"` to leave single elements of record-of values undefined. As usual, read access to undefined fields, comparing only partially defined record-of values or sending of only partially defined record-of values will cause a test case error.

Access to the individual elements of a record-of value is achieved with the subscription operator `"[]"` with indices starting from 0. Writing elements past the current end of a record-of does not cause a test case error, as long as it does not conflict with any length restriction placed on the type. Instead, the record-of value is extended to accommodate the additional elements. Table 8.24 shows how this can be exploited to implement concatenation of two record-of values. The latter example also shows how the number of elements of a record-of value can be accessed with the `lengthof` function. Note that when the `lengthof` operation is used with lists it returns the sequential number of the last initialised element.

```
lengthof( {1,2,3,4} )          // evaluates to 4
lengthof( {6,6,6,-, 6} )       // evaluates to 5
```

8.4.5.2   Arrays

In TTCN-3, as in many programming languages, it is possible to use arrays to group values. Arrays can be defined either inline in constant or variable declarations, or they can be defined as types in their own right. Examples of both are shown in Table 8.25. The number of elements in an array is specified between square brackets. Each array must specify the number of elements that the array is going to contain. When only a single number n is given, the values in the array are indexed from 0 to $n - 1$. It is also possible to specify upper and lower bounds for the indices. The bounds must both be constant, positive values of type `integer`. The upper and lower bound always belong to the index range.

Table 8.26 shows examples of initialisation and assignment for arrays. When specifying a value for an entire array, the value list notation is used. Note that such an assignment for an array must specify the exact number of elements declared in the array definition. Like for record-of values, it is possible to leave certain elements undefined by using the `"-"` symbol. Such values may then be specified later, but reading uninitialised array elements or using partially defined arrays in assignment or comparisons will cause a test case error. As shown in the examples the pre-defined function `lengthof` can also be used for arrays. As the index of the first element in an array can be different to zero, the

Table 8.23 Defining record-of types and values

```
type record length ( 1..infinity ) of ViaHeader ViaHeaders;

type record ViaHeader {
   charstring sentProtocol,
   charstring sentBy,
   charstring viaParams optional
};

const ViaHeaders c_aliceViaHeaders :=
{ {   sentBy      := "SIP/2.0/UDP",
      sentBy      := "bigbox3.site3.atlanta.com",
      viaParams   := omit
  },
  {   sentBy      := "SIP/2.0/UDP",
      sentBy      := "pc33.atlanta.com",
      viaParams   := omit
  }
};

const ViaHeaders c_switchedViaHeaders :=
{ {   sentBy      := "SIP/2.0/UDP",
      sentBy      := "pc33.atlanta.com",
      viaParams   := omit
  },
  {   sentBy      := "SIP/2.0/UDP",
      sentBy      := "bigbox3.site3.atlanta.com",
      viaParams   := omit
  }
};

// since order different
if (c_aliceViaHeaders == c_switchedViaHeaders ) ...      // evaluates to false

// but after element rotation
if (c_aliceViaHeaders == (c_switchedViaHeaders @> 1)) ...
     // evaluates to true


const ViaHeaders c_moreProxies :=
{ {   sentBy      := "SIP/2.0/UDP",
      sentBy      := "bigbox1.site1.atlanta.com",
      viaParams   := omit
  },
{ {   sentBy      := "SIP/2.0/UDP",
      sentBy      := "bigbox2.site2.atlanta.com",
      viaParams   := omit
  }
}
// example record-of value concatentation
const ViaHeaders c_aliceViaHeadersLong := c_moreProxies & c_aliceViaHeaders;

type record length ( 4 ) of unsignedbyte Ipv4Address;
```

Table 8.23 (*continued*)

```
const IPv4Address c_localHost := { 127, 0, 0, 1};
lengthof(c_aliceViaHeaders )    // returns 2
replace(c_localHost, 0, 2, {128, 55} )    // returns {128, 55, 0, 1}
```

Table 8.24 Element access for record-of value

```
var record of charstring v_masterProxyList;
/*... */
var record of charstring v_aliceProxyList := {c_aliceViaHeaders[0].sentby,
                                              c_aliceViaHeaders[1].sentby };

function f_append( inout record of charstring p_listA,
                   in    record of charstring p_listB ) {
  var integer v_sizeA := lengthof( p_listA );
  var integer v_sizeB := lengthof( p_listB );
  var integer i;

  for ( i := 0; i < v_sizeB; i := i + 1 ) {
    p_listA[v_sizeA + i] := p_listB[i];
  }
};
/*... */
f_append( v_masterProxyList, v_aliceProxyList)
```

Table 8.25 Defining and using arrays

```
type unsignedshort IPV6Address[8];  // explicit array definition, 8x16 bit
var unsignedbyte v_ipv4Address[4];  // implicit array definition, 4x8 bit

var IPV6Address v_ipv6Address := {   // using the explicit definition
  65152, 255, 42554, 17, 0, 0, 35, 1
};

v_ipv4Address[0] := 127;            // accessing single array elements
v_ipv4Address[1] := 0;
v_ipv4Address[2] := 0;
v_ipv4Address[3] := 1;

const SipStatusCode c_scTrying       := 100;
const SipStatusCode c_scNotAcceptable := 606;
var charstring v_reasonPhrases[c_scTrying .. c_scNotAcceptable];
```

amount of elements in an array is determined by the index of the last element minus 1 and minus the index of the first element.

As shown in the previous examples, access to array elements is achieved with the subscription operator "[]". The index specified must be between the array's bounds. TTCN-3 provides array bound checking so that access to elements outside the declared range of an array is caught as a test case error.

Table 8.26   Partially defined arrays

```
// ERROR - 4 elements required
var unsignedbyte v_ipv4Addr[4]        := {0,0,0};


// OK, two elements undefined
var unsignedbyte v_localAddr[4]       := {192, 15, -, -};

const unsignedbyte c_myLocalAddr[4] := {192, 15, 17, 42};

// OK, second element is defined
var unsignedbyte v_localAddr[4]       := {-, 15, 0, -};
lengthof( v_localAddr )          // evaluates to 2!
const unsignedbyte c_myLocalAddr[4] := {192, 15, 17, 42};
lengthof( c_myLocalAddr )     // evaluates to 4

v_localAddr[0] := 192;
v_localAddr[3] := 100;

if ( v_localAddr == c_myLocalAddr ) ...          // OK now
```

8.4.5.3   Multi-Dimensional Arrays

TTCN-3 also allows the definition of multi-dimensional arrays. These can be used to store multi-dimensional tables or matrices. A small example for the multiplication of a 2-by-3 matrix with a 3-by-2 matrix is given in Table 8.27. A number of details in this example are worth pointing out:

- Multi-dimensional arrays are defined by adding *one range per dimension* after the value or type name like this TwoByThree[0..1][0..2]. The corresponding syntax is used to access individual array elements.
- When specifying values for multi-dimensional arrays, the **left-most** index range specifies the **outer-most** dimension of the value. Note how the value c_2x3 consists of an array with two elements, each of which contains three elements.
- Multi-dimensional arrays are treated as nested one-dimensional arrays. This makes it possible to project a multi-dimensional array to a less-dimensional array. This is used to extract the nth vector from p_2x3, which is then assigned to an array that can store three integer values – this works because the type TwoByThree is treated as an array that can hold two elements, each of which is a three-element array.

Unlike all other defined TTCN-3 types, there exists no way to define subtypes for arrays.[1] Indeed, arrays in TTCN-3 are not a particularly well-developed concept and in most cases it is better advised to use record-of types instead of arrays. Record-of types provide the same functionality as arrays, but are better integrated into TTCN-3. For any collection of values that will be used for communication with the SUT, record-of or set-of

---

[1] This is a peculiarity of TTCN-3 that results from the fact that, initially, TTCN-3 was not supposed to allow the definition of explicit array types at all.

Table 8.27 Matrix multiplication using arrays

```
type integer TwoByThree[2][3];
type integer ThreeByTwo[0..2][0..1];
type integer TwoByTwo[2][2];

const TwoByThree c_2x3 := { {1,2,3}, {4,5,6} };
const ThreeByTwo c_3x2 := { {1,2}, {3,4}, {5,6} };

function f_multiply( in TwoByThree p_2x3, in ThreeByTwo p_3x2 )
return TwoByTwo {
  var TwoByTwo v_result := { {0,0}, {0,0} };
  var integer n,m,i;

  for ( n := 0; n < 2; n := n + 1 ) {
    var integer v_row[3] := p_2x3[n];      // extracting vector from p_2x3
    for ( m := 0; m < 2; m := m + 1 ) {
      for ( i := 0; i < 3; i := i + 1 ) {
        v_result[n][m] := v_result[n][m] + v_row[i] * p_3x2[i][m];
      }
    }
  }
  return v_result;
}

// f_multiply( c_2x3, c_3x2 ) evaluates to {{22,28}, {49,64}};
```

types should be used. The only advantage of arrays over record-of types is that arrays do not require an explicit definition and hence provide a lightweight means to define auxiliary list variables when they are needed. It can also be useful when the list's elements cannot be naturally numbered from 0, like in the reasonPhrases example from Table 8.25. Record-of values are necessarily indexed from 0.

8.4.5.4 The Set-of Type

Sometimes, the strict ordering of fields, as is imposed by a record-of type or array, does not adequately reflect the way that data is treated by the tested application or protocol. An example from the SIP context is the Allow header, which is used to indicate which SIP requests a communication party is prepared to accept. These methods are listed in the header in an arbitrary order, so it would not be natural to impose an order in the TTCN-3 representation.

Set-of types are similar to record-of types in nearly every aspect. The value notation is the same. Access using the subscription operator "[]" is possible and indeed treats a set-of value as if it were a record-of value. Its elements can be rotated and different set-of values can be concatenated. Also, the lengthof function is available to determine the length of a set-of value. An example of a set-of type definition for the SIP Allow header is shown in Table 8.28.

The only notable difference between set-of and record-of is the notion of equality, which disregards the order of elements in set-of values. While two record-of values are equal if

Table 8.28   Defining set-of types and values

```
type charstring SipMethod ( "REGISTER", "INVITE", "ACK", "BYE",
                            "CANCEL", "OPTIONS" );

type set of SipMethod AllowHeader;

const AllowHeader c_allowedMethods := {
  "INVITE", "ACK", "BYE", "CANCEL", "OPTIONS"
};
```

they contain the same values *in the same order*, two set-of values are already considered equal if they contain the same elements in the same multiplicity, but not necessarily in the same order. For example, the following two constants are equivalent, although elements with the same index are different.

```
const AllowHeader c_elementaryMethods := { "INVITE", "ACK",
                                           "BYE" };
const AllowHeader c_supportedMethods  := { "BYE", "ACK",
                                           "INVITE" };

if ( c_elementaryMethods == c_supportedMethods ) ...
  // evaluates to true
```

Set-of values are not real sets in the mathematical sense but rather multi-sets, that is sets that also keep track of the multiplicity of their elements, so that {1,2,3,1} and {3,1,1,2} are equal but {1,1,2,3} and {1,2,3} are not.

8.5   Nested Type Definitions

A fairly recent addition to the TTCN-3 language is the ability to defined nested type definitions for record, set, record of, set of and union types – a concept, that is also available in other notations such as ASN.1 [39]. As a matter of fact this concept was mainly introduced to better support the import of data structures defined in ASN.1. Nested type definitions allow types including all of the structured types or subtype restrictions to be specified without giving these field types explicit identifiers as shown in Table 8.29.

Note that the return values of pre-defined functions lengthof and sizeof on values of nested type definitions do not take elements of structured fields into account. Although this concept can reduce the amount of TTCN-3 code written for type definitions and makes sense to use for list types and subtyping fields of basic type, we do not advise this concept is used for the definition or subtyping of structured fields for record, set and union types, since it reduces the readability of code and error notifications, and more importantly prevents reusability of field type definitions.

Table 8.29   Example nested type definitions

```
type record of charstring AllowHeaders; // simple nested type definition
const AllowHeaders c_AllowHeaders := { "PRACK", "Event", "Subscription" };
const charstring c_firstAllowedMethod := c_supportedMethods[0]; // = PRACK

type record (1..infinity) of record { // more complex nested type definition
        charstring        sentProtocol,
        HostPort          sentBy,
        GenericParams     viaParams optional // type can be re-used elsewhere
} ViaHeaders;

type record (1..infinity) of record { // valid definition but not
                                       // recommended
        charstring        sentProtocol,
        HostPort          sentBy,
        record ( 1..infinity ) of record { // type cannot be re-used
                                            // elsewhere!
            charstring id,
            charstring pValue optional
        } viaParams optional
} ViaHeaders;

type record CSeqHeader {
  integer number (0 .. infinity), // nested subtype definition
  Method  method
};
```

8.6   Encoding and Decoding of Data

As we mentioned already in the introduction of this book, prior to sending structured TTCN-3 data values to the SUT they are encoded by the codecs in a test system. Similarly, encoded messages are automatically decoded into TTCN-3 data structures after they have been received by the test system. The invocation of these codec functions is usually performed implicitly when a send or receive statement is executed. In some cases, however, it may be useful to access encoded values in TTCN-3 code. For this purpose the predefined functions encvalue and decvalue can be used.

Let us go back to our SIP example: As shown in our introduction to this protocol in Table 8.1 SIP messages may contain a message body which essentially is a payload. Since the SIP standard allows anything that can be expressed in the form of characters to be attached in the message body part, one way of defining a general SIP message type is to define the message body a field of type charstring in a record type. The SDP is one possible payload that can be transported in SIP messages. This text-based protocol is used to exchange information about voice codecs available in the terminal of the caller and callee.

In Table 8.30 we show a simplified definition of a SDP message. In addition, we show how the pre-defined function encvalue, which itself returns only a bitstring value,

Table 8.30   Example use of encvalue and decvalue

```
type record SdpMessage { // simplified SDP message definition
    charstring origin,
    charstring connection optional,
    set of charstring mediaDescriptions optional,
    set of charstring attributes optional
}

template SdpMessage a_exampleSdp := {
    origin := "53655765 2353687637 IN IP4 pc33.atlanta.com",
    connection := "IN IP4 pc33.atlanta.com",
    mediaDescriptions:= {"audio 3456 RTP/AVP 0 1 3 99"},
    attributes := {"rtpmap:0 PCMU/8000"}
}
const charstring c_encExampleSdp :=
        oct2char(bit2oct(encvalue(a_exampleSdp)));

var SdpMessage v_decodedSdp;
if ( decvalue(oct2bit(char2oct(c_encExampleSdp)),
             v_decodedSdp) > 0 ) {
  // in this case value of v_decodedSdp remains undefined
  log( "Could not decode the following as SDP:",
       c_encExampleSdp );
} else {
  match (v_decodedSdp, a_exampleSdp) // returns true
}
```

can be used to encode a structured SDP message value or template resolving to a specific value. In a SIP test, we could of course assign this value straight to the message body field of the SIP request. The `decvalue` function operates in a similar manner but requires arguments, next to the encoded value in `bitstring` format, also a reference to a variable of the type to which the encoded message shall be attempted to be decoded to. The return value of `decvalue` indicates the success of the decoding.

8.7   Summary

Now that we have studied the various ways to use and define types in TTCN-3, defining a type for (slightly simplified) SIP messages is relatively simple, by building on the numerous type examples that we have encountered in this chapter. In addition, we have learned about a number of different pre-defined functions for conversion between different types, encoding and decoding of values. It should be mentioned that our presented type definitions are just *one* possible way to define parts of SIP or SDP messages in TTCN-3.

Indeed, some choices that we have made when modelling the types were motivated by the wish to cover all of TTCN-3's type system.

With this chapter, we have covered the most important aspects of the TTCN-3 type system, which enable the type representations for most application domains. There are a few, more advanced concepts that we have so far left out and which will be covered in the following chapter.

9

Advanced Type Topics

In our presentation of the Testing and Test Control Notation Version 3 (TTCN-3) type system in the previous chapter, some aspects have been left out deliberately to allow for a more stringent presentation of the most important topics. However, these aspects are sufficiently fundamental (like type compatibility) or useful for particular applications (like the `anytype`, recursive type definitions and foreign type system support) that this book would be incomplete without them. These aspects will now be covered in this chapter.

9.1 Type Compatibility

Like many other programming languages, TTCN-3 is a statically typed language. This means that each entity in TTCN-3 is declared together with its type. For a variable, its type describes which values can be stored in the variable and how it can be used to form expressions. For a function, it is specified what values can be passed as parameters, and what values can be returned as return values. For a port, it is defined, which values can be sent and received via this port and so on. For each entity, its type is known statically from the program code. It can hence be used to catch ill-formed expressions. In the following, we will describe the rules used to determine whether an expression is well typed. These rules guide which assignments, expressions and parameter instantiations are considered type-safe.

We have already discussed TTCN-3's capabilities to define new types from existing types via subtyping (see Section 8.2). As it turns out, subtyping and type compatibility are closely related concepts in TTCN-3. Subtype definitions must be acyclic; this guarantees that each user-defined subtype has a uniquely determined *root type*. This *root type* is the built-in or user-defined structured type, from which the type is ultimately derived by the chain of subtype definitions. For example, `integer` is the root type of all three types `SipStatusCode`, `SipContentLength` and `SipInfoStatusCode` in the following definitions.

An Introduction to TTCN-3, Second Edition.
Colin Willcock, Thomas Deiß, Stephan Tobies, Stefan Keil, Federico Engler and Stephan Schulz.

```
type integer SipStatusCode ( 100 .. 609 );
type SipStatusCode SipInfoStatusCode ( 100, 180..183 );
type integer SipContentLength ( 0 .. infinity );
```

Two types in TTCN-3 are *compatible* if they have the same root type. So, for example `SipStatusCode`, `SipInfoStatusCode`, `SipContentLength` and `integer` are mutually compatible; they all have `integer` as their root type. On the other hand, `SipStatusCode` and `charstring` are **not** compatible, since they have different root types (`integer` vs `charstring`). This is all there is to say about type compatibility for types derived from built-in types, but there are some more rules for inter-type compatibility for structured types, which we will briefly mention later.

TTCN-3 is a statically typed language, which means that each object – each variable, constant, literal, template, function, test case or altstep – has a declared type. This type governs which values an object may assume, which operations may be applied to the object, how objects may be combined with operators to compound expressions, and which objects may be used in the instantiation of parameterised entities.

The type compatibility rules of TTCN-3 allow objects of compatible type to be used interchangeably. The only exceptions to this rule are the communication statements, where a different notion of type compatibility is used (see Section 9.1.1). Each occurrence of type incompatibility is statically detectable and will be flagged as a type error. Table 9.1 shows examples of correctly and incorrectly typed expressions.

We can identify two different classes of type errors in Table 9.1: static *type incompatibility*, like comparing two entities of incompatible types or passing arguments of incompatible types, and *subtype violations* where values of compatible types are used outside subtype restrictions required by the context.

Both kinds are type errors, but only the type incompatibilities are always statically detectable and can hence be flagged by a TTCN-3 tool before the execution of the test suite. Subtype violations may only be detectable during test suite execution. For example, it will only be detectable at run time that `f_fib( 100 )` is of astronomical size and exceeds by far `255`. Depending on your TTCN-3 tool, some of these situations will be detected statically, like the illegal call `f_fib( -1 )`, but no tool can detect all possible violations statically. Those that are only detected during run time will then cause test case errors. This is not necessarily bad and should not prevent you from using subtype restrictions whenever possible. Calling the `f_fib` function with a negative argument would, if not caught as a subtype violation, cause an infinite recursion and thus probably an uncontrolled crash of the test system execution because of memory exhaustion. The restriction of the argument of `f_fib` to be of type `Byte`, and hence positive, will catch such erroneous invocation directly as a subtype violation.

9.1.1   Strict Type Compatibility

There is one class of situations in which the type compatibility rules that we have just described do not apply: the `send`- and `receive`-like statements `send`, `call`, `raise`, `receive`, `getcall`, `catch` and `trigger`. The normal rules of type compatibility do not apply for the templates used as arguments to these operations and the optional value redirections. Instead, a template or variable may only be used in one of these statements

Table 9.1 Correctly and incorrectly typed expressions

```
type integer PositiveInt ( 0 .. infinity );
type integer Byte ( 0 .. 255 );

function f_fib( in Byte p_n ) return PositiveInt {
   var PositiveInt x := 1;
   var Byte smallResult := 1;
   // OK: comparison of integer with Byte
   if ( p_n == 0 ) {
     return 1;
   // TYPE ERROR: cannot compare Byte with float
   } else if ( p_n == 0.5 ) {
   // TYPE ERROR: cannot return float value as
   //            PositiveInt
     return 1.0;
   // OK: comparison of Byte with PositiveInt
   } else if ( p_n == x ) {
     return smallResult;                  // OK: return Byte as PositiveInt
   } else {
     return fib( p_n-1 ) + fib( p_n-2 ); // OK: substract integer from
                                          //     Byte, add PositiveInt
   }
}

  function f_foo() {
    // OK: pass integer as Byte, store ( small )
    //     PositiveInt result in Byte
    var Byte result1 := fib( 10 );
    // TYPE ERROR: passing float value for
    //             Byte parameter
    var Byte result2 := fib( 10.0 );
    // DYNAMIC TYPE ERROR: fib( 100 ) > 255;
    var Byte result3 := fib( 100 );
    var PositiveInt result3 := fib( -1 );  // TYPE ERROR: -1 violates Byte's
                                           //             restriction
    // DYNAMIC TYPE ERROR: result -9
    // violates PositiveInt's restriction
    var PositiveInt result4 := fib( 1 )  -  10;
    var float result5 := fib( 10 ) / 2.0;  // TYPE ERROR: cannot divide
                                           //             PositiveInt by float
  }
```

if, and only if, its exact type is explicitly specified in the communication port's type for the correct communication direction.

Practically, this means that for a send operation on a port whose type specifies integer as its only out type, only integer templates may be used for sending. Using subtypes and even aliases of integer constitutes a type error. Similarly, if the port specifies only a specific subtype of integer as its out type, integer templates may not be sent. Examples are shown in Table 9.2. The reason for TTCN-3 to be so restrictive in send- and receive-like statements is that these statements are directly concerned with

Table 9.2 Strict type compatibility for communication operations

```
type integer IntegralNumber;

type port NumberPort message {
    out integer;
    in IntegralNumber;
}

template integer a_fortytwo := 42;
template integer a_someInt := ?;
template IntegralNumber a_seventeen := 17;
template IntegralNumber a_someIN    := ?;

// assume port pt is of type NumberPort

pt.send( a_fortytwo );      // OK
pt.send( a_seventeen );     // TYPE ERROR
pt.receive( a_someInt );    // TYPE ERROR
pt.receive( a_someIN );     // OK

var integer v_int;
var IntegralNumber v_IN;

pt.receive( a_someIN ) -> value v_IN;   // OK
// TYPE ERROR: v_int is no IntegralNumber var
pt.receive( a_someIN ) -> value v_int;
```

communication, and hence with encoding and decoding of values. In TTCN-3, encoders and decoders are tied to specific types. This allows the selection of different encodings of values based, for example on the different user-defined aliases of integer (see also Section 8.2.1). So, in the above example, the encoder used for integer may use a different encoding than that for IntegralNumber values.

Explicit type annotations must be used to distinguish implicit templates whose types cannot otherwise be inferred from contextual information or to cast expressions to compatible types to be used for sending and receiving of values. Table 9.3 shows examples for this. Implicit templates for real message types are best avoided because they are hard to read and cannot be reused elsewhere in the code.

9.1.2   Type Compatibility for Structured and Special Types

So far, we have introduced only a single rule for type compatibility: two types are compatible if they have the same root type, that is if they are derived via chains of subtype restrictions from the same built-in or structured type. For types derived from built-in types, that is indeed all there is to type compatibility. However, for structured types, there are a few additional rules, when two types with different root types are compatible. Since there are very few situations where this form of type compatibility will improve readability, or

Table 9.3 Using explicit types to disambiguate implicit templates

```
type integer IntegralNumber;

type port AmbiguousPort message {
    inout integer, IntegralNumber;
};



// assume port pt is of type AmbiguousPort


pt.receive( ? );                    // TYPE ERROR: ambiguity of ?

pt.receive( IntegralNumber:? ); // OK
pt.send( 42 );                      // TYPE ERROR: ambiguity of 42
pt.send( integer:42 );              // OK
```

maintainability of your code, we do not treat it here. If necessary, refer to the TTCN-3 core language standard [8] for further information.

The situation is different for special types – in particular for component types. Here, type compatibility can be very useful to increase the reuse of function definitions. Assuming strict type compatibility we would not be able to call functions from a function or testcase statement with different component type references in their runs on clause. But runs on clauses do not follow such strict type compatibility rules. As a matter of fact a function or testcase statement can call another function as long as their referenced component types are compatible, that is the body of the component type associated with the invoked function is an exact subset of the one associated with the invoking function or test case. This is particularly the case for components that are the result of component type extension as described in Section 5.4. An example is show in Table 9.4.

9.2   The Anytype Type

The TTCN-3 core language does not allow types to be used as parameters,[1] like, for example Abstract Syntax Notation One (ASN.1) and to a limited extent C++ and Java do. At the same time, TTCN-3 has very strict rules for type compatibility and does not allow free conversion between types like many scripting languages or C (via casting to void pointers) do. With the lack of these capabilities, the implementation of generic protocols or algorithms would be impossible if TTCN-3 did not have the special type anytype.

The type anytype is a built-in union type that, in each module, contains one variant for each type visible in that module. We will see how the anytype can be used for similar purposes as type parameterisation, but first, we explain how anytype values can be specified and manipulated. The anytype type is used just like any other union type, with the exception that it does not need to be defined. The anytype is always implicitly

---

[1] Type parameterisation is defined in a separate TTCN-3 extension standard which is discussed in Section 12.3.

Table 9.4 Exploiting component type compatibility for function reuse and invocation

```
type port StringPort message { inout charsting };

type component StringComp { port StringPort pt_str };
type component TimerComp { timer t_time };

// extension means that TimedStringComp has port pt_str and timer t
type component TimedStringComp extends StringComp {
  timer t;
};

function f_sendHelloWorld() runs on StringComp {
  pt_str.send("Hello World!");
};
function f_startTimer() runs on TimerComp {
  t_time.start(5.0);
};

testcase tc_timedHelloWorld() runs on TimedStringComp {
  f_startTimer();    // ERROR - TimeComp and TimedStringComp use different
                     //          timer identifiers (t versus t_time)!
  f_sendHelloWorld(); // OK - StringComp and TimedStringComp are compatible
  alt {
     [] pt_str.receive { setverdict(pass); t.stop; }
     [] t.timeout  { setverdict(fail);
  }
  f_send42();   //
}
```

Table 9.5 Using the anytype

```
var anytype x   := { integer := 42 };
var integer x1 := x.integer;
x.charstring    := "Now I am a charstring value";
x.float         := 10.0;
var integer y  := x.integer;              // ERROR  -  float selected
var integer y1 := float2int( x.float );   // OK  -  explicit conversion
                                          // required
```

defined in each module and is indeed a different type for each module. The name for each variant of the anytype is the name of the type of that variant. Like ordinary union types in TTCN-3, the anytype is a *tagged* union, which means that only the selected variant can be read from a value. In other words, the anytype **cannot** be used to cast values between different types, instead the predefined conversion functions must be used for this purpose. Examples for the use of anytype can be found in Table 9.5.

Without qualification, the type of an anytype value is anytype, regardless of its selected variant. Any setting and reading of an anytype value (with anything else but anytype values) must explicitly specify the type variant, see Table 9.6.

Table 9.6   Correct and incorrect access to values stored in an `anytype` value

```
var anytype x_any  := { charstring := "some charstring value" };
var charstring str := x_any;                                    // ERROR
var anytype x_str  := x_any.charstring;                         // ERROR
var anytype x_any1 := x_any;                                    // OK
```

Table 9.7   Using the `anytype` in the presence of imported types

```
module mod_a {
    type integer A;
    type float D;
}

module mod_b {
    type integer B;
    type float D;
}

module mod_c {
    type integer C;
    type integer D;
    type integer E;
}

module mod_d {
    import from mod_a all;
    import from mod_b all;

    type integer E;

    const anytype c_1 := { A := 1 }; // from mod_a
    const anytype c_2 := { B := 2 }; // from mod_b
    const anytype c_3 := { C := 3 }; // ERROR - C is not known in mod_d
    // ERROR - unqualified D is not visible in mod_d
    const anytype c_4 := { D := 4 };
    const anytype c_5 := { mod_a.D := 2.0 };
    const anytype c_6 := { mod_b.D := c_5.mod_a.D };
    const anytype c_7 := { E := 1 }; // locally defined E
}
```

It is important to remember that the `anytype` is defined locally to each module and it contains exactly those types visible in that module. For names that have been defined more than once (via imports), the usual visibility rules apply, as defined in Chapter 7. Some examples for what that means when referring to variants of an `anytype` are shown in Table 9.7. Note how the access to the D variant (from module mod_a) in the declaration of c_6 uses the module prefix mod_a to distinguish between the different declarations of D visible in mod_d.

9.2.1 Using the anytype for Generic Protocol Definitions and Data Types

One important usage of the anytype is to simulate parameterised types that are, for example available through templates in C++, or are commonly available in functional programming languages. This is particularly useful when testing involves some form of transport protocol, which should also be modelled within your test system, for example when testing more than one level of a protocol stack simultaneously. In this case, the protocol definition of the transport protocol should ideally be done independently of the protocol's payload type. One way to do this is to model the payload as a binary or character string type. Unfortunately, this would require explicit en- and decoding of the payload within the TTCN-3 code. Another way is to use anytype for the payload, which allows moving the en- and decoding into the codec implementation, that is outside of the TTCN-3 code.

The first of these approaches is adequate as long as the payload does not play an important role in the testing. However, if the information, that is carried in the payload needs to be inspected more closely, a representation as plain charstring is no longer sufficient because it makes the generation of data to be sent and the inspection of received values unnecessarily complicated. A more structured data type would help in this process.

In the SIP example this brings us to the question as to which type should be chosen to represent the SIP payload. If only a single kind of data will be carried by SIP messages, we would of course choose a representation of that data for the payload type. A suitable union could be used if a fixed number of known types were transported by SIP messages. But in each of these cases, the definition of SIP explicitly references the payload types. How can the transport protocol be defined without explicit references to the payload? The answer is the anytype, which allows the transport protocol to be defined independently from the payload. A first attempt to define an abstract transport protocol with a generic payload is given in Table 9.8. We will discuss the shortcomings of this approach and refine it further in the remainder of this section.

We have already hinted that this is not the final solution to the problem, so what is the problem with this definition? The problem is the anytype, which, as we have described, is always a union over all types *visible* in the current module. In our case, this would mean that the ProtocolDataUnit could carry exactly those types that are visible in the module TransportProtocol, including imported types. This means that each payload type has to be either defined in the module TransportProtocol or at least has to be imported into that module. Adding a new payload requires direct modification of TransportProtocol, which is undesirable.

Table 9.8 A generic transport protocol, first attempt

```
module transport {
    type ProtocolDataUnit {
      ProtocolHeader header,
      anytype          payload
    }
}
```

Table 9.9 A generic transport protocol, second attempt

```
module payload {
    // add definitions or imports for all payload types here

    type anytype Payload;
}

module transport {
    import from payload { type Payload };

    type ProtocolDataUnit {
      ProtocolHeader header,
      Payload        payload_
    }
}
```

We cannot completely avoid this problem, but we can at least localise the modification to a dedicated position, using a slightly more complex module structure to define the transport protocol. This is shown in Table 9.9, where a separate module is used to define a dedicated `Payload` type to be used in the definition of `ProtocolDataUnit`. While this does not decouple the definition of `ProtocolDataUnit` completely from the payload, it is probably as close to a real generic definition as you can get in TTCN-3. It does not require much imagination to see how this approach could, for example be used to define generic data structures, like container types.

Type parameterisation, see Section 12.3, provides a more consistent way of leaving types open in other definitions.

9.3 The Address Type

The SUT may consist of several entities that may have to be addressed individually in communication operations. Coming back to our SIP example, certain test scenarios might involve communication with several communication partners. Depending on the modelling of the communication with the SUT, this will involve addressing several individual entities that communicate via a single port at the test system interface. This single test system interface port, channels all TCP or UDP communication from the test system to the SUT.

To support the handling of addresses inside the SUT, TTCN-3 offers the special type `address`. This type may be set to the literal `null` to specify undefined address values. Values and variables of type `address` may be used with the `to`, `from` and `sender` clauses in communication statements, if the port in question is currently mapped to a test system interface port. Examples are shown in Table 9.10.

Of course, the actual data representation of `address` values will be highly dependent on the application domain. For SIP, an address type combining IP addresses plus TCP or UDP port numbers could be used. A different example would be CORBA-based systems, where object references are used to communicate with the typically numerous objects

Table 9.10 Using `address` in sender redirection and addressing

```
altstep alt_alwaysBusy( SipPort pt_sip ) {
    var address v_callerAddress := null;

    [] pt_sip.receive( SipRequest : a_anyInviteRequest )
          -> sender v_callerAddress {
        pt_sip.send( SipResponse : a_busyResponse ) to v_callerAddress;
      }
}
```

that comprise the SUT. Here, interchangeable object references (IORs) could be used as address values.

Apart from the different structure of these address domains, there is another important difference between IP addresses and IORs. An IP address can easily be specified *externally* by a user who knows the IP address and listening port number of a running SIP server; an IP address is transparent and meaningful to the user. For CORBA-based systems, this is different: valid addresses are usually obtained either from a trading service or other CORBA components, in other words, valid addresses are usually created from *within* the SUT; their structure is largely meaningless and opaque to the user. TTCN-3 caters for the application-dependent nature of addresses and allows appropriate representation of both transparent and opaque addresses.

In the case of transparent addresses, it is possible to give a definition for the value representation of the `address` type within TTCN-3. This definition must be unique for the whole module, that is there may be at most one specification of the data representation for the `address` type per module. For our SIP example, an appropriate definition for the `address` type is given in Table 9.11. When the data representation of `address` has been specified, it is possible to create new address values within the TTCN-3 code.

For the case of opaque addresses, it is also possible to leave the data representation open within the TTCN-3 code. This is done by simply **not** giving a definition for `address`. In that case, values of type `address` will be treated opaquely in the TTCN-3; it is not possible to create new address values in TTCN-3 (other than `null`) and all addresses have to be obtained via communication with the SUT. The actual data representation is left to the codec and System Adaptor (SA) implementation.

9.4 Recursive Type Definitions

We have already seen how user-defined types may be used in the definition of new user-defined types to form more and more complex data structures. This has been thoroughly discussed in Section 8.4. What we have not mentioned in this section is that it is even possible to define types in a recursive manner, with the type definition referring to itself, possibly via a number of type definition steps. This feature can be useful in modelling dynamic data structures like trees or other complex data structures. Special care has to be taken to make sure that instances of these recursive types remain finite, that is infinite recursion is avoided.

In this section, we will have a close look at recursive type definitions. We do this by discussing different type definitions that can be used to represent binary trees in

Table 9.11   Using IP addresses for the address type

```
type record length ( 4 ) of unsignedbyte IPv4Address;
type record length ( 8 ) of unsignedshort IPv6Address

type union IPAddress {
    IPv4Address ipv4,
    IPv6Address ipv6
};

type record Address {
    IPAddress      ipAddress;
    unsignedshort portNumber
};

type Address address;              // fix data representation of address
const address c_localProxy := {
    ipAddress  := { ipv4 := {127,0,0,1}},
    portNumber := 5060
};

/* ... */
pt.send( a_sipRequest ) to c_localProxy;
```

Table 9.12   A recursive type definition for binary trees, first attempt

```
type record InfiniteBinaryTreeBad {
    integer element,
    InfiniteBinaryTreeBad leftSubTree, // ERROR - infinite recursion
    InfiniteBinaryTreeBad rightSubTree // ERROR - infinite recursion
};
type record InfiniteBinaryTree {
    integer element,
    InfiniteBinaryTree leftSubTree optional, // OK - recursion is resolvable
    InfiniteBinaryTree rightSubTree optional // OK - recursion is resolvable
};

var InfiniteBinaryTree c_tree := {
    element := 1,
    leftSubTree := {
      element   := 0
      omit // no subtree
    },
    omit // no right subtree
};
```

TTCN-3. We will highlight the potential benefits and shortcomings of different modelling approaches and thus introduce you to the important ideas of recursive type definitions. The first attempt at such a type definition is given in Table 9.12. In structural type definitions cyclic definitions are not allowed. To ensure that no infinite recursion occurs, record and set types can only define optional fields recursively as shown in our example. For union types this is achieved by requiring that at least one alternative is not recursively defined. The type InfiniteBinaryTree is a perfectly valid type and is close to how one would define a tree structure in other programming languages like C or Pascal, with the one difference that these languages would use pointers to InfiniteBinaryTree for the sub trees.[2] In these languages, distinct null pointers can be used to indicate a node has no left or right sub-tree. TTCN-3 does not support pointers but allows the use of omit, which could be exploited to specify a value of InfiniteBinaryTree.

The approach with optional fields suffers from the fact that there is no simple way to represent the empty tree since omit by itself is not a valid value in TTCN-3. This is a severe shortcoming when trying to express tree traversal functions in a recursive manner, which is the natural approach when processing recursive data structures. A better definition of a binary tree is to use a union type that enables exiting the type recursion. An example of such a union type is shown in Table 9.13.

Table 9.13 Using unions to define recursive types with finite values

```
type boolean Null; // any type really would work here

type union BinaryTree {
    BinaryTree_union tree,
    Null null_
};

type record BinaryTree_union {
    integer     element,
    BinaryTree leftSubTree,
    BinaryTree rightSubTree
};

const BinaryTree c_empty := { null_ := false };

const BinaryTree c_tree := {
    tree := {
      element := 1,
      leftSubTree :=
      { tree :=
        {
          element := 0,
          leftSubTree  := c_empty,
          rightSubTree := c_empty
        }
      },
      rightSubTree := c_empty  }
};
```

---

[2] In these languages, it is indeed *necessary* to define the subtrees in terms of pointers because they do not allow direct recursion with type definitions.

Table 9.14 Recursively calculating the sum of node values in a `BinaryTree`

```
function f_sumOfElements( in BinaryTree p_tree ) return integer {
    if ( p_tree == c_empty ) {                     // exit recursion
                                                   // for empty trees

      return 0;
    }
    else {
      return p_tree.tree.element +
              f_sumOfElements( p_tree.tree.leftSubTree ) +
              f_sumOfElements( p_tree.tree.rightSubTree )
    }
};
```

Table 9.15 Creating cyclic data structures with recursive types

```
var BinaryTree v_cycleTree := {
    tree := {
      element := 0,
      leftSubTree  := c_empty,
      rightSubTree := c_empty
    }
};

v_cycleTree.tree.leftSubTree  := v_cycleTree;
v_cycleTree.tree.rightSubTree := v_cycleTree;
```

In the modelling from Table 9.13, the empty tree is represented by the value `c_empty`, which allows for simple tree traversal, as shown in Table 9.14. You can see how the possibility to use the empty tree directly as a value allows for an elegant expression of the algorithm.

To wrap up this section, we come back to the issue of cyclic data structures. Table 9.15 shows the creation of a tree that folds back on itself. Cyclic structures like this could, for example be used to represent a ring buffer. They may also be useful to create pathological input data for software testing.

9.5 Foreign Type Systems

The richness of TTCN-3's type system has been a recurring theme in the last chapters. Using the large range of the TTCN-3 type system, it is possible to find close representations of the SUT data format within TTCN-3. For some SUTs, it is even possible to use the SUT data type definitions directly, even though they will, of course, not normally be specified in TTCN-3. Standardised mappings from common data description languages to the TTCN-3 type system make this possible. Currently, such mappings exist for data type definitions in the ASN.1 [39], OMG's Interface Definition Language (IDL)

[27, 34] and for XML schemata [15]. ASN.1 is strongly rooted in the telecommunication protocol world, and, for example many of the protocols that form the Universal Mobile Telephony System (UMTS) protocol stacks are specified in ASN.1. IDL has its origins in the world of CORBA and is mainly used to describe the interfaces of distributed, object-oriented systems.

The degree of integration of foreign type systems into TTCN-3 will depend on the TTCN-3 tool, that is used. The support for these languages ranges from the possibility to directly use these foreign types inside TTCN-3 to the automatic generation of standardised codecs and system adaptation.

9.5.1   Using ASN.1 Types in TTCN-3

The integration of ASN.1 into TTCN-3 follows the approach that ASN.1 modules can be imported *directly* into TTCN-3 modules. This means that the types and values defined in the imported ASN.1 module can then be used directly as if they were defined in any normal imported TTCN-3 module. This approach is possible because the type systems of TTCN-3 and ASN.1 are sufficiently close and there exists an obvious mapping between the types and values within the two languages. As an example, consider the ASN.1 module from Table 9.16. This module contains data type definitions for a simplified version of the Domain Name System (DNS) protocol. Note that we use a different set of type definitions for DNS than the one that has been previously used in this book. The new modelling shown here is closer to the actual message structures described in the DNS specification [29].

Using this ASN.1 module in TTCN-3 is as simple as importing it into a TTCN-3 module, as shown in Table 9.17. Note the attribute language "ASN.1:2002" is used to specify that we are importing from a module that contains ASN.1 definitions conforming to the ASN.1 standard from 2002 [39]. Other ASN.1 standards are also supported – refer to [20] for details.

The result of this import is the same as if the TTCN-3 module from Table 9.18 had been imported. All the types and constants from the ASN.1 module DNS can be used within the imported module as if they were imported from the module DNS_TTCN. For example, these imported definitions can be used in the definition of constants, new structured types, and they can be further restricted by subtypes and so on.

There are a number of subtleties to observe when importing ASN.1 modules into TTCN-3, though. Most importantly, ASN.1 and TTCN-3 have different syntactic rules. ASN.1 allows hyphens ' - ' to be used within identifiers, for example within Table 9.16 the definition of the type DNS-Message. In TTCN-3, this is not allowed. This problem is circumvented by implicitly turning each hyphen into an underscore ' _ ' during the import of ASN.1, so that the type DNS-Message becomes accessible in TTCN-3 as DNS_Message. Another problem is the difference in the reserved keywords of ASN.1 and TTCN-3. In our example, the definition of Type in Table 9.16 uses the TTCN-3 keyword any, and the definition of Answer uses the keyword type. Such clashes are resolved by appending the underscore character to each such identifier when importing the module. The module DNS_TTCN shows the result of such suffixing.

Finally, when importing ASN.1 into TTCN-3, the set of ASN.1 keywords must also not be used in TTCN-3; this means that it is not possible to refer to ASN.1 built-in types

Table 9.16   Data type definitions for DNS in ASN.1

```
DNS
DEFINITIONS AUTOMATIC TAGS ::=
BEGIN

    UnsignedByte  ::= INTEGER ( 0 .. 255 )
    UnsignedShort ::= INTEGER ( 0 .. 65536 )
    IPv4Address   ::= SEQUENCE ( SIZE ( 4 ) ) OF UnsignedByte
    Name  ::= SEQUENCE OF Label              -- dotted names are broken up
    Label ::= IA5String ( SIZE ( 0 .. 63 ) )  -- into a sequence of labels

    Type  ::= ENUMERATED                     -- kinds of data available via
                                             -- DNS
    {
      a( 1 ), ns( 2 ), cname( 5 ), mx( 15 ), any( 255 )
      -- truncated: more types available
    }

    Rdata ::= CHOICE                         -- resource records that
                                             -- contain DNS data
    {
     a        IPv4Address,
     cname    IA5String,
     ns       IA5String
     -- truncated: more choices available, e.g., arbitrary text records
    }

    Query ::= SEQUENCE                       -- a request to retrieve info
                                             -- from DNS
    {
      qname Name,
      qtype Type
      -- truncated: also contains a class field
    }

    Answer ::= SEQUENCE                      -- an answer to a query
    {
      name    Name,
      type    Type,
      rdata   Rdata
      -- truncated: also contains time to live information
    }

    Answers     ::= SEQUENCE OF Answer
    Queries     ::= SEQUENCE OF Query

    DNS-Message ::= SEQUENCE                 -- the messages exchanged
    {                                        -- between DNS clients and
                                             -- servers
      id        UnsignedShort,
      queries   Queries,
      answers   Answers
      -- truncated: also contains flags, authorative and additional info
    }
END
```

Table 9.17 Importing ASN.1 modules into TTCN-3

```
module dns_test {
    import from DNS language "ASN.1:2002" all;

    // more imports and definitions here
}
```

Table 9.18 TTCN-3's equivalent to the ASN.1 module from Table 9.16

```
module DNS_TTCN {
    type integer UnsignedByte ( 0 .. 255 );
    type integer UnsignedShort ( 0 .. 65536 );

    type record length ( 4 ) of UnsignedByte IPv4Address;

    type record of Label Name;
    type charstring Label length ( 0..63 );

    type enumerated Type {
      a( 1 ), ns( 2 ), cname( 5 ), mx( 15 ), any( 255 )
    }

    type union Rdata {
      IPv4Address a,
      charstring  cname,
      charstring  ns
    }

    type record Query {
      Name qname,
      Type qtype
    }

    type record Answer {
      Name    name,
      Type    type_,
      Rdata rdata
    }

    type record of Answer Answers;
    type record of Query Queries;

    type record DNS_Message {
      UnsignedShort id,
      Queries        queries,
      Answers        answers
    }
}
```

Table 9.19 Correspondence between ASN.1 type as TTCN-3 types

ASN.1 type	Maps to TTCN-3 equivalent
BOOLEAN	**boolean**
INTEGER	**integer**
REAL[a]	**float**
OBJECT IDENTIFIER	**objid**
BIT STRING	**bitstring**
OCTET STRING	**octetstring**
SEQUENCE	**record**
SEQUENCE OF	**record of**
SET	**set**
SET OF	**set of**
ENUMERATED	**enumerated**
CHOICE	**union**
VisibleString	**charstring**
IA5String	**charstring**
UniversalString	**universal charstring**

[a]For the subtleties caused by the specification of bases other than 10 for ASN.1 REAL types, refer to [8, Annex D].

directly. Instead, type or value aliases can be defined in the ASN.1 modules, which are then usable also within TTCN-3. For an example, see Section 9.5.1.2.

9.5.1.1 Correspondence between ASN.1 and TTCN-3 Types

In the example above, we have already seen that most ASN.1 types are mapped directly to their TTCN-3 counterparts. Specifically, the Table 9.19 shows the ASN.1 types that can be mapped directly, and their TTCN-3 equivalent. Note that the TTCN-3 type `objid` is not part of the TTCN-3 core language, it is introduced specifically to TTCN-3 in part 7 of the standard [20].

9.5.1.2 The ASN.1 NULL Type

ASN.1 has a built-in type called NULL, which has only a single value, also called NULL. NULL is used extensively in ASN.1 protocol definitions, but a direct equivalent type in TTCN-3 does not exist. The type NULL may occur in an ASN.1 module either in the definition of a synonym type or within a larger type definition. In the first case, the synonym type is imported to TTCN-3 as an enumerated type with a single element NULL. In the second case, a nested type definition is used, again replacing the ASN.1 type NULL with an enumerated type as before.

We have already mentioned that no ASN.1 keyword must be used as a TTCN-3 identifier when ASN.1 is used with TTCN-3 including the identifier NULL. This makes it impossible to use the value NULL within TTCN-3. To circumvent this problem, it is often convenient to define an alias for the ASN.1 type and value NULL within an ASN.1 module, as is shown in Table 9.20.

Table 9.20 Using NULL type and value in TTCN-3

```
-- define a synonym name for the NULL value

AsnNull       ::= NULL
asnNull NULL ::= NULL

-- This can be used in TTCN-3 now as follows:
-- const AsnNull c_null := asnNull
```

9.5.1.3 Open Types, Subtyping and Parameterisation

ASN.1 allows type definitions that use open types, that is types that are not further specified within the ASN.1 modules. In TTCN-3, type parameters allow the definition of open types in a similar way as explained in Section 12.3. Nevertheless, such open types in ASN.1 cannot be imported. Only types with value parameters can be imported to TTCN-3.

The example from Tables 9.16 and 9.18 already shows that certain forms of subtyping defined in the ASN.1 module are preserved when imported into TTCN-3. As a rule of thumb, those subtype constraints that can be expressed in TTCN-3, for example restrictions to particular values, value ranges or size restrictions, are also valid when the ASN.1 module is used from within TTCN-3. Others, like user-defined constraints, will be ignored when the ASN.1 is used within TTCN-3. For a complete list of how ASN.1 subtyping is treated within TTCN-3, refer to [20].

9.5.2 Using IDL Types in TTCN-3

Like for ASN.1, a mapping from IDL to TTCN-3 has been defined to allow direct use of IDL definitions from within TTCN-3 [14]. Unlike ASN.1, standardisation has chosen an indirect mapping approach: instead of directly importing IDL modules into TTCN-3, a real TTCN-3 module has to be generated by an IDL-to-TTCN-3 compiler. This generated module can then be imported by other TTCN-3 modules. The standard document [14] provides the details of this mapping.

IDL has its roots in the object-oriented world where (remote) method invocation serves as the primary means of communication. This means that in those places where object-orientation is used in IDL, the mapping has to be more complicated to bridge the gap between IDL and the non-object-oriented TTCN-3. The IDL structured types and enumerated types are converted to TTCN-3 in the expected, straightforward manner. Naturally, TTCN-3 procedure-based communication is used to model the (remote) method invocation of IDL. Indeed, the procedure-based communication mechanism of TTCN-3, with the separation of in, out and inout parameters, has been modelled on IDL to serve exactly this purpose. Finally, methods that are collected by IDL interfaces are converted into TTCN-3 port definitions. These ports can then be used to communicate with the instances of the implemented IDL interfaces in the SUT.

Chapter 6 used an IDL example of a simple directory server to expose important ideas of procedure-based communication in TTCN-3. This example is repeated in Table 9.21.

Table 9.21   IDL definitions for a simple directory service

```
module DirectoryService {

    exception NotAllowed { string reason; };
    exception Rejected { string reason; };
    exception KeyNotFound { };
    exception SessionExpired { };

    enum Capabilities {
      e_reader, e_readerwriter
    };

    interface Directory {
      string lookup( in string key )
        raises ( KeyNotFound, SessionExpired );

      void update( in string key, inout string val )
        raises ( NotAllowed, SessionExpired );
      void logout();
    };

    interface DirectoryManager {
      Directory login( in string username,
                       in string password,
                       out Capabilities capabilities )
        raises ( Rejected );
    };
};
```

The (slightly simplified) result of the translation according to the mapping rules from [14] is shown in Table 9.22. The interfaces `Directory` and `DirectoryManager` are mapped to port types and the TTCN-3 type `address` is used to pass object references as return values of the `login` method. Note, that each signature has an exception of type `SYSTEM_EXCEPTION`. This type is a predefined type of the IDL to TTCN-3 mapping, it is a union on the system exceptions from IDL, for example NO_PERMISSION to indicate that the caller does not have sufficient rights to perform the operation. We will not go into further detail about this mapping, refer to [14] for further information.

9.5.3 Mapping XML to TTCN-3

Many applications use XML to map their data into a standardised format that can be used over networks and on different platforms. To test XML-based protocols or applications, it is necessary to map XML data types into a format that can be used with TTCN-3. A natural starting point for this are XML Schemas, which describe the specific syntax of XML data used by the SUT. Currently, there are two alternatives to use XML data types in TTCN-3:

Table 9.22   The result for mapping the DirectoryService IDL file to TTCN-3

```
module DirectoryService {

    type record NotAllowed { charstring reason }
    type record Rejected { charstring reason }
    type record KeyNotFound { }
    type record SessionExpired { }

    type enumerated Capabilities {
      e_reader, e_readerwriter
    }

    signature lookup ( in charstring key ) return charstring
      exception ( KeyNotFound, SessionExpired, SYSTEM_EXCEPTION );

    signature update ( in charstring key, inout charstring val )
      exception ( NotAllowed, SessionExpired, SYSTEM_EXCEPTION );

    signature logout()exception ( SYSTEM_EXCEPTION );

    type port Directory procedure {
      inout lookup;
      inout update;
      inout logout;
    }

    type address DirectoryObject;

    signature login ( in charstring username, in charstring password,
                      out Capabilities capabilities ) return
                      DirectoryObject
      exception ( Rejected, SYSTEM_EXCEPTION );
    type port DirectoryManager procedure {
        inout login;
    }

    type address DirectoryManagerObject;
}
```

The *implicit* way uses the **language** and **import** keywords and requires the definition of a specific data interface containing all structural and encoding information. Typically, this is realised directly by the TTCN-3 tool and cannot be altered by the user. The *explicit* mapping of data types translates the XML schema definitions (XSDs) directly into corresponding TTCN-3 data types. These mappings, as well as the encoding information, are accessible by the user.

The transformation of data types from XSD to TTCN-3 is performed in two steps: first the XSD is transformed into ASN.1 according to ITU-T RecommendationX.694 [40] and thereafter into TTCN-3 using the standard conversion rules described in Section 9.5.1 [20]. Additionally, specific extensions are used to retain the original XSD

Table 9.23 XML phonebook Schema

```
<xs:schema xmlns:xs="http://www.w3.org/2001/XMLSchema">
   <xs:element name="phonebookentry">
      <xs:complexType>
         <xs:sequence>
            <xs:attribute name="city" type="xs:string" use="required"/>
            <xs:element name="person" type="xs:string"/>
            <xs:element name="address">
               <xs:complexType>
                  <xs:sequence>
                     <xs:element name="lastname" type="xs:string"/>
                     <xs:element name="firstname" type="xs:string"/>
                     <xs:element name="street" type="xs:string"/>
                  </xs:sequence>
               </xs:complexType>
            </xs:element>
            <xs:element name="phonenumbers">
               <xs:complexType>
                  <xs:sequence>
                     <xs:element name="fixed" type="xs:long"/>
                     <xs:element name="mobile" type="xs:long"
                                             minOccurs="0"/>
                  </xs:sequence>
               </xs:complexType>
            </xs:element>

         </xs:sequence>
      </xs:complexType>
   </xs:element>
</xs:schema>
```

nature of a given TTCN-3 type. Note, some TTCN-3 tool may choose to perform the transformation in a single step, however the outcome must be the same as in the two-step approach.

As an example of implicit use of XSD data types, consider the simple XML Schema, shown in Table 9.23, that describes the structure of phonebook entries. Now, we will see how this schema is translated into corresponding TTCN-3 types and a TTCN-3 template translated back into XML data.

Conversion of the XML schema to TTCN-3 (via ASN.1) yields the type definitions shown in Table 9.24. Note how the 'minOccurs= "0"' in the XML yields optional fields in the TTCN-3 for the mobile number, and how the information about the city being an attribute is added as a variant statement in the TTCN-3 type. Lastly note how the address field in the XML is renamed to address_ in the TTCN-3 to avoid a conflict with the associated keyword.

Using these type definitions, we can define the template shown in Table 9.25. This template can then be encoded back into XML format as shown in Table 9.26.

The exact mapping rules are described in part 9 of the TTCN-3 standard [15] and we will not go into further detail in this book.

Table 9.24 XML phonebook Schema converted into TTCN-3

```
module Example1 {
  import from XSD language "XML" all;
  type record PhonebookEntry {
    XSD.String city,
    XSD.String person,
    record
    {
      XSD.String lastname,
      XSD.String firstname,
      XSD.String street
    } address_,
    record
    {
      XSD.long fixed,
      XSD.long mobile optional
    } phonenumbers
  } with {
      variant "lastname as uncapitalized";
      variant(address_.street)"name as 'homeaddress'";
      variant(city) "attribute";
  }
} with {
  encode "XML";
}
```

Table 9.25 Example of a TTCN-3 template using XML types

```
module XmplTemplate {
  import from XSD language "XML" all;
  import from Example1 all;

  template PhonebookEntry t_PhonebookEntry:={
    city:="Honolul",
    person:="Dr.Steinhuber",

    address_:=
    {
      lastname:="Steinhuber",
      firstname:="Martin",
      street:="Pestalozzistreet 22a"
    },

    phonenumbers:=
    {
      fixed:=2312212332,
      mobile:= omit
    }
  }
}//end module
```

Table 9.26 The encoding of the phonebook entry template

```
<?xml version="1.0" encoding="UTF-8"?>
<phonebookentry city="Honolulu">
<person>"Dr.Steinhuber"<person>
<address>
 <lastname>Steinhuber<\lastname>
 <firstname>Martin<\firstname>
 <homeaddress>"Pestalozzistreet 22a"<\homeaddress>
<\address>
<phonenumbers>
 <fixed>2312212332<\fixed>
<\phonenumbers>
<\phonebookentry>
```

9.6 Summary

This concludes our discussion of the more advanced aspects of the TTCN-3 type system. Together with Chapter 8, we have now given a near complete introduction to the types of TTCN-3 and outlined how the different built-in and user-defined structured types can be used to find appropriate representations of the data from your application domain. This is particularly easy when the application domain provides type definitions in a type system that can be directly used from TTCN-3. In these cases, much of the work of mapping the application data to TTCN-3 is already done by the standardised mappings that exist for these languages.

In the other cases where no such mapping exists, more work is needed from your side. Yet, with the rich set of built-in and structured types, plus the expressive power of recursive type definitions and the `anytype`, it should be possible to find good representations of the application data.

10

Templates

The exchange of information via messages plays a prominent role in the black box testing of protocols, but is also important in the testing of systems that employ various forms of (remote) procedure call interaction. To help users in efficiently defining the information to be sent and received in such communication, Testing and Test Control Notation Version 3 (TTCN-3) allows the definition of so-called templates. Templates are closely coupled to TTCN-3 types. TTCN-3 types define the logical structure of messages, whereas templates specify their information content.

The simplest form of TTCN-3 templates defines a unique value, which is mainly used as the argument to sending operations. The real power of TTCN-3 templates, however, lies in their ability to specify multiple values or variations of a message within one single definition, which can then be used in a `receive` statement. This feature allows the user to handle complex messages by only focusing on the relevant parts and ignoring information, which is not of interest in a given test situation. After discussing how this selective matching can be specified, we will go on to introduce regular expressions, which is one of the important new additions to the TTCN-3 language.

We have already briefly introduced templates in our discussion of the different communication operations in Chapters 4–6. In the next two chapters, however, we are less concerned with the actual usage of templates but rather with the different mechanisms that exist to define templates in TTCN-3 and how templates and template matching are handled during run time.

10.1 A First Look at TTCN-3 Templates

Templates must always be defined on the basis of a type and with a template identifier. Template definitions, in their simplest form, look very similar to TTCN-3 constant definitions with the keyword `const` replaced by `template` as shown in Table 10.1. In these cases, template definitions define one single value of the type that they reference. The template identifier prefix 'a_' used in our examples is a naming convention that we follow in this book (see Chapter 15). It should be noted that the TTCN-3 language itself does not require the use of such a convention.

An Introduction to TTCN-3, Second Edition.
Colin Willcock, Thomas Deiß, Stephan Tobies, Stefan Keil, Federico Engler and Stephan Schulz.
© 2011 John Wiley & Sons, Ltd. Published 2011 by John Wiley & Sons, Ltd.

Table 10.1   Example template definitions

```
template charstring a_register    := "REGISTER";
template HostPort    a_nokiaAt5060 := {
  host        := "nokia.com",
  portNumber := 5060
}
```

Table 10.2   The definition of the simplified `HostPort` type

```
type record HostPort {
  charstring host,                                // IP Address or domain
                                                  // name
  integer    portNumber ( 0 .. 65565 ) optional  // optional port
}
```

The type definition underlying the template `a_nokiaAt5060` is shown in Table 10.2. Here, the type of the `host` field has been changed from `IpAddress`, which has been used in the previous chapters, to `charstring` in order to simplify some of the following examples in this chapter.

Like regular TTCN-3 values, template definitions can be written using either an assignment syntax or a list syntax. Our previous template definition presented in Table 10.1 used the assignment syntax. This could equally be specified with the list syntax as follows.

```
template HostPort a_nokiaAt5060 := {"nokia.com", 5060}
```

In this chapter, we will mainly use the more verbose form to increase the readability of our template definitions.

Template definitions may also contain constant references, constant expressions, module parameter references and even function invocations. Examples of this are shown in Table 10.3.

Table 10.3   Template definitions with constant reference and constant expressions

```
const float c_pi    := 3.1415927;
modulepar { charstring mp_host := "foo.fi" };

template float      a_pi    := c_pi;

template charstring a_CRLF := int2char( 10 ) & int2char( 13 );

template HostPort a_hostAt5060 := {
  // will only be "foo.fi" if not initialised externally
  host        := mp_host,
  portNumber := 5060
}
```

Table 10.4   Some simple example template definitions with matching expressions

```
template integer  a_anyContentLength        := ?; // any integer is ok
template HostPort a_nokiaOrFooAtAnyOrNoPort := {
  host       := ( "nokia.com", "foo.fi" ), // one of the two is ok
  portNumber := *                          // any or no integer is ok
}
```

Often in testing, when we receive a message we are only interested in checking certain specific fields and values. This means much of the information in such messages is not of immediate interest. By using templates in receiving operations, TTCN-3 allows such parts to be ignored by using matching expressions. The next example, shown in Table 10.4, defines the template a_anyContentLength. This template allows any `integer` value to be received. The following template definition a_nokiaOrFooAtAnyOrNoPort accepts two possible host names and any or no port number. These examples show how we can concentrate on specifically testing the values of certain fields, while at the same time only broadly checking other message elements.

10.2   The TTCN-3 Match Operation

Templates are tightly coupled to one of the most powerful features of the TTCN-3 language: its built-in matching mechanism. During test case execution, whenever a test system receives a message, this matching mechanism (which is implemented by the TTCN-3 tool) checks the incoming value against the expected message definition, specified by the template.

This matching mechanism, which is automatically invoked when a TTCN-3 `receive` operation is executed, can also be directly invoked using the TTCN-3 `match` operation. The `match` operation takes two parameters: the first is a value and the second is a template. The operation checks if the given value is within the restrictions given by the template. It returns `true` in the case of a match, otherwise it returns `false`. The `match` operation may also be called with a value and template of different types. In this case, the operation always returns `false`. The result of the match operation can also be logged via the `log` statement.

Table 10.5   An example use of the TTCN-3 `match` operation

```
const HostPort c_nokiaAt5060 := {
  host := "nokia.com", portNumber   := 5060
}

const HostPort c_nokiaNoPort := {
  host := "nokia.com", portNumber := omit
}

// the a_nokiaAt5060 template was defined in Table 10-1
b := match( c_nokiaAt5060, a_nokiaAt5060 ); // b evaluates to true
// b evaluates to false, omit != 5060
b := match( c_nokiaNoPort, a_nokiaAt5060 );
```

Table 10.5 shows an example of matching a `record` value against the template `a_nokiaAt5060`, which has been defined in Table 10.1. The template successfully matches the value `c_nokiaAt5060` because both fields of the `record` template match with the ones in the constant value. The template does not match the value `c_nokiaNoPort` because the template specifies a port number, whereas the constant does not.

In the remainder of this chapter, we will use the `match` operation to illustrate the effect of different forms of template specification on the matching mechanism.

10.3 Template Definition for One Specific Value

As discussed before, a template in its simplest form can be thought of as a TTCN-3 constant value. Such a template matches a value if, and only if, the value and template are identical. Template definitions with a unique value can be specified for any type, that is for built-in types, structured types and subtypes. The rule for such template definitions is that the template only defines one specific legal value of that type.

The previous sections have already given a number of examples for such templates. Table 10.6 provides some more example definitions. The template `a_binaryMsgBody` shows that in the case of `hexstring` and `octetstring` hexadecimal digits are treated independently of their case by the matching mechanism. The second `match` operation for

Table 10.6   Example template definitions for one specific value

```
// example template of basic type
template integer a_statusCodeOK := 200;
b := match( 100, a_statusCodeOK ); // b evaluates to false, 100 != 200
b := match( 200, a_statusCodeOK ); // b evaluates to true

// example string template
template octetstring a_binaryMsgBody := 'Cafe'O;
b := match( 'cafe'O, a_binaryMsgBody ); // b evaluates to true
b := match( 'CAFE'O, a_binaryMsgBody ); // b evaluates to true
// 'cafe'O != 'caff'O
b := match( 'caff'O, a_binaryMsgBody ); // b evaluates to false,

// example union template
union IpAddress {
  charstring ipv4,
  charstring ipv6,
  charstring hostName
}

template IpAddress a_nokiaHostName := { hostName := "nokia.com" };
const IpAddress c_loopIpv4         := { ipv4     := "127.0.0.1" };
const IpAddress c_nokiaHostName    := { hostName := "nokia.com" };
b := match( c_nokiaHostName, a_nokiaHostName ); // b evaluates to true
// variant different
b := match( c_loopIpv4,      a_nokiaHostName ); // b evaluates to false,
```

the union template yields `false` because in `c_loopIpv4` a different variant is selected than that in the template.

Such specific value templates *must* be used in sending operations since the value of the messages we send to the System Under Test (SUT) must be unambiguous. Such templates are also used for receiving operations.

10.4 Template Definitions with Matching Expressions

In receiving operations, it is often unnecessary and sometimes even impossible to specify the exact message content, that is expected. In TTCN-3, this can be handled within the template definition by replacing explicit values by a matching expression. TTCN-3 offers a number of matching expressions to deal with such cases. These expressions can be used to specify that multiple values or even any value are acceptable from the SUT for a given field or set of fields. Such template definitions using matching expressions may never be used in sending operations.

The use of matching expression in template definitions is closely linked to the underlying type of the template. Since there are many matching mechanisms and a large type system to cover, we will first present in this section the more generally applicable matching expressions. We will then continue in the following sections to introduce new and extend already introduced matching expressions in the context of specific types. In the summary of this chapter, we will finally provide a comprehensive overview of all the matching expressions and their relation to TTCN-3 types.

10.4.1 The 'any' Matching Expression

'any' (denoted by '?') is probably the most frequently used matching expression. It can be applied to any built-in type, string type or user-defined type. This expression accepts any single value, which is compatible with the underlying type definition. Should the underlying type definition be restricted in its values because of subtyping, the 'any' matching expression will not match those values that do not comply with this subtype restriction.

In the examples shown in Table 10.7, the `integer` type `UInt` is subtyped to non-negative values, therefore, the template `a_anyUInt` will not match the value `-1`. The next example shows that the empty string value is also included when the 'any' expression is applied to a string value. It should be noted that in the same manner the empty list is also included in case of `record of` and `set of` values. The `IpAddress` values used by our last example have been defined in the previous table.

10.4.2 Value Lists

When specifying expected values, the use of the 'any' matching expression may sometimes be too liberal. In such cases, we may only want to allow a handful of alternatives, that is a value list. A template definition with a value list simply specifies all the values that are acceptable. As well as specific values, it is also possible to refer to other template definitions in the value list. A received value will match the value list if it matches one

Table 10.7 Example template definitions with the 'any' matching expression

```
// example template of built-in type
type integer UInt ( 0 .. infinity ); // non-negative numbers
template integer a_anyInteger := ?;
template UInt    a_anyUInt    := ?;

b := match(   1, a_anyUInt   ); // b evaluates to true
// -1 not within type range
b := match(  -1, a_anyUInt   ); // b evaluates to false,
b := match(  -1, a_anyInteger ); // b evaluates to true

// example template of string type
template charstring a_anyTextString := ?;
b := match( "",          a_anyTextString ); // b evaluates to true
b := match( "TTCN-3",    a_anyTextString ); // b evaluates to true
// bitstring != charstring
b := match( '101110'B, a_anyTextString ); // b evaluates to false,

// example template of user defined type ( union )
template IpAddress a_anyIpAddress := ?;
b := match( c_nokiaHostName, a_anyIpAddress ); // b evaluates to true
b := match( c_loopIpv4,      a_anyIpAddress ); // b evaluates to true
```

of the elements in the list. Like for specific values, all the values and templates specified in a value list must comply with subtyping restrictions of the underlying type, if such restrictions exist.

Table 10.8 shows some example template definitions of built-in, string and user-defined types. Here, the template a_nokiaOrFooAtPort shows how a value list may also be used for user-defined values.

10.4.3 Complemented Value List

In some cases, it may be easier or shorter to specify in a template definition the values a field shall not take rather than the values that it shall. For this purpose, the complement matching expression can be used. A value will only match such a template definition if it is *not* equal to the values or constants, which are given as an argument to that expression. It is also possible to refer to other templates inside a complemented values list. Examples of templates with the complement matching expression are shown in Table 10.9.

10.4.4 Value Ranges

Template definitions of the basic types integer and float may be defined as a range of acceptable values. A value will match such a template if it is between the lower and upper limit of the range. The range can be defined with one or both boundary values included or excluded from the range. To exclude a boundary value from the range it has to be preceded by the character '!'. The pre-defined constants infinity and −infinity can be used if no upper or lower limit shall be specified for the expected value. As our

Table 10.8 Example template definitions with value lists

```
// example template of built-in type
template integer a_twoInfoStatusCodes := ( c_tryingCode, 180 );
const integer c_tryingCode := 100;

b := match( 100, a_twoInfoStatusCodes ); // b evaluates to true
b := match( 180, a_twoInfoStatusCodes ); // b evaluates to true
// 200 != 100,180
b := match( 200, a_twoInfoStatusCodes ); // b evaluates to false,

// example template of string type
template charstring a_basicMethod :=
         ( "REGISTER","INVITE","ACK","BYE","CANCEL","OPTIONS" );

b := match( "BYE",    a_basicMethod ); // b evaluates to true
// "NOTIFY" not in list
b := match( "NOTIFY", a_basicMethod ); // b evaluates to false,

// example template of user-defined type ( record )
template HostPort a_nokiaOrFooAtPort:= (
  { host      := "nokia.com", portNumber := 5060 },
  { host := "foo.fi", portNumber    := 5070 }
)
const HostPort c_fooAt8080 := {
  host := "foo.fi",
  portNumber := 8080
}

b := match( c_nokiaAt5060, a_nokiaOrFooAtPort ); // b evaluates to true
// 8080 != 5070
b := match( c_fooAt8080,   a_nokiaOrFooAtPort ); // b evaluates to false,
```

Table 10.9 Example template definition with a complemented value list

```
template charstring a_illegalSipVersion := complement( "SIP/2.0" );
b := match( "SIP/1.8", a_illegalSipVersion ); // b evaluates to true
// "SIP/2.0" == "SIP/2.0"
b := match( "SIP/2.0", a_illegalSipVersion ); // b evaluates to false,

// templates can also be used in a complement expression
template charstring a_legalSipVersion := complement( a_illegalSipVersion );
```

example template a_anyInfoStatusCode in Table 10.10 shows, value ranges may also be used within value list templates.

In this example, the mismatch between −1 and the template a_unsignedInteger is not due to the violation of a subtype restriction. The value −1 is a valid integer and thus among the values of the type definition underlying a_unsignedInteger. It is a mismatch because it is simply not among the expected values specified by the template definition.

Table 10.10 Example template definitions with a value range

```
template integer a_unsignedInteger    := ( 0 .. infinity );
template integer a_anyInfoStatusCode := ( 100, 180 .. 183 );

b := match(  -1, a_unsignedInteger   ); // b evaluates to false
b := match( 183, a_anyInfoStatusCode ); // b evaluates to true
// 110 != 100,180..183
b := match( 110, a_anyInfoStatusCode ); // b evaluates to false,
```

10.4.5 More about Matching Expression for Structured Types

Now that we have the knowledge of the most fundamental matching mechanisms, we will introduce new matching expressions, which are only applicable for structured types, that is record, set and union types. But first we will briefly revisit the 'any' matching expression in the specific context of structured types.

10.4.5.1 The 'any' Matching Expression within a Structured Type

When specifying a template for a structured type, the 'any' matching expression can also be applied to the fields of the structured type, that is *within* a structured value. This use allows some parts of a structured type in a template definition to be matched against a specific value while others parts are left unrestricted.

When used for an optional field in a record or set type, '?' will only match when a field value is present. This is shown by the template a_nokiaAtAnyPort in Table 10.11. This template does not match c_nokiaAtAnyPort, where the portNumber fields have been omitted. The template a_anyIpv4Address shows how 'any' can be used within a union value to allow any value for a specific variant. In contrast, the template a_anyIpAddress shows the use of 'any', which is independent of the present variant.

10.4.5.2 The 'any-or-none' Matching Expression

As we have discussed in the previous section, there are cases where we may want to completely ignore the state of an optional field in a set or record value, that is we would like to accept any value as well as the case where the field is absent. Table 10.12 shows how the 'any-or-none' matching expression (denoted by '*') allows the specification of such a constraint for an optional field. Our new template definition a_nokiaHostAnyOrNoPort now matches any specific value, as well as the absence of the portNumber field.

10.4.5.3 Allowing Omission

When specifying a template for record or set types, it may be desirable to express that an optional field should either contain no value, that is omit, or one specific value. Similarly, we may want to accept a list or range of values and omit. The

Table 10.11 User-defined template definitions with any expression for fields

```
// template example expecting a record value
template HostPort a_nokiaAtAnyPort := {
  host       := "nokia.com",
  portNumber := ?
}

const HostPort c_nokiaAt5060 := {
  host := "nokia.com", portNumber := 5060
}

const HostPort c_nokiaNoPort := {
  host := "nokia.com", portNumber := omit
}

b := match( c_nokiaAt5060, a_nokiaAtAnyPort ); // b evaluates to true
b := match( c_nokiaNoPort, a_nokiaAtAnyPort ); // b evaluates to false

// template example expecting a union value
template IpAddress a_anyIpv4Address := { ipv4 := ? }
template IpAddress a_anyIpAddress   := ?;

const IpAddress c_loopIpv4       := { ipv4     := "127.0.0.1" };
const IpAddress c_nokiaHostName := { hostName := "nokia.com" };
b := match( c_loopIpv4,       a_anyIpv4Address ); // b evaluates to true
// variant hostName != ipv4
b := match( c_nokiaHostName, a_anyIpv4Address ); // b evaluates to false
b := match( c_nokiaHostName, a_anyIpAddress );   // b evaluates to true
```

Table 10.12 Example template definitions with the 'any-or-none' expression

```
template HostPort a_nokiaAtAnyOrNoPort := {
  host       := "nokia.com",
  portNumber := *
}
// b evaluates to true
b := match( c_hostNoPort,       a_nokiaHostAnyOrNoPort );
// b evaluates to true
b := match( c_hostAnd5060Port, a_nokiaHostAnyOrNoPort );
```

ifpresent matching attribute, which is added after the expected value(s), can be used to specify such a constraint. Note that placing omit directly in the list or range of expected values is not allowed. In Table 10.13, the first two values match the template a_nokiaAt506XorNoPort (see Table 10.12) because their value of the field port-Number is either omitted or within the specified value range. In contrast, the value c_nokiaAt8080 causes a mismatch since the field value is neither omitted nor within the value range.

Table 10.13 Example template definition with `ifpresent` matching attribute

```
const HostPort c_nokiaAt8080 := {
  host       := "nokia.com",
  portNumber := 8080
}

template HostPort a_nokiaAt506XorNoPort:= {
  host       := "nokia.com",
  portNumber := ( 5060 .. 5069 ) ifpresent
}

b := match( c_nokiaNoPort, a_nokiaAt506XorNoPort ); // b evaluates to true
b := match( c_nokiaAt5060, a_nokiaAt506XorNoPort ); // b evaluates to true
// 8080 != 5060,omit
b := match( c_nokiaAt8080, a_nokiaAt506XorNoPort ); // b evaluates to false
```

Table 10.14 Example template definition to `omit` a field

```
template HostPort a_nokiaNoPort := {
  host       := "nokia.com",
  portNumber := omit
}

b := match( c_nokiaNoPort, a_nokiaNoPort ); // b evaluates to true
// 5060 != omit
b := match( c_nokiaAt5060, a_nokiaNoPort ); // b evaluates to false,
```

10.4.5.4 Requiring Omission

It can also be described in a template that an optional field of a `record` or `set` type *must* be omitted. In this case, the keyword `omit` should be assigned to the field. In our example in Table 10.14, the value `c_nokiaAt5060` does not match the template because it provides a value for the field `portNumber`.

10.4.6 More about Matching Expressions for List-Like Types

We will introduce here four additional matching expressions `length`, `permutation`, `superset` and `subset` that TTCN-3 offers specifically for list values. But first we will revisit the 'any' and 'any-or-none' matching expressions in the context of these specific values.

10.4.6.1 The 'any' and 'any-or-none' Matching Expressions within List-Like Types

When used *within* `record of` or `set of` templates, the 'any' matching expression can be used instead of a specific list element value. Similarly, as in the case within structured types, this expression will match any list element as long as it is present.

Table 10.15 Using 'any-or-none' in `record of` templates

```
// example of record of type
type record of integer IntSequence;

// template examples expecting for a record of value
template IntSequence a_containsOneTwo  := { *, 1, 2, * };
template IntSequence a_4IntsWithOneTwo := { ?, 1, 2, ? };

b := match( { 1,2 },      a_containsOneTwo  ); // b evaluates to true
// wrong order of 1 and 2
b := match( { 2,1 },      a_containsOneTwo  ); // b evaluates to false
b := match( { 3,1,2,1,2 }, a_containsOneTwo  ); // b evaluates to true
// 1,3 != 1,2
b := match( { 1,3,2 },    a_containsOneTwo  ); // b evaluates to false,
// 2,1 != 1,2
b := match( { 1,2,1,2 },  a_4IntsWithOneTwo ); // b evaluates to false,
```

The 'any-or-none' matching expression, however, takes on a different meaning. In this case, the expression does not include the omitted value, since there is no such thing as `optional` in list types. Instead, 'any-or-none' matches any *number* of (additional) list elements as well as the empty list value. These expressions give users a powerful tool to check for the occurrence of specific list elements especially in large lists.

Table 10.15 shows examples of using 'any' as well as 'any-or-none' in `record of` templates. The template `a_containsOneTwo` matches any sequence of integers that contains the sequence 1, 2. It also matches the value {1,2} because it does not require additional preceding and trailing integers. It does not match the value {1,3,2} because the value 3 occurs in between 1 and 2. The template `a_4IntsWithOneTwo` shows that 'any' contrary to 'any-or-none' requires one integer value prior to the sequence.

The evaluation of `set of` templates requires much more work for the matching mechanism. A `set of` value matches with a template if its list elements match any permutation of the list element values or matching expressions specified by the template. Table 10.16 shows examples of `set of` templates. The templates `a_oneAndAnother` require exactly two elements for a match. Therefore, they do not match the values {1} and {3,2,1}. However, it matches the value {2,1} since the order of elements is not important in a `set of` type. The template `a_containsOne` matches any value that contains a 1 as well as any additional integers. Since 'any-or-none' also accepts zero additional elements, the `set of` value {1} also matches. Because of the unordered nature of a `set of` type, the position of both expressions as well as multiple occurrences of 'any-or-none' are not relevant within a `set of` template.

10.4.6.2 Length Restrictions of List-Like Types

It is possible to restrict the length of the expected list value using the `length` matching attribute. The argument of this attribute may either be a single integer or an integer range, possibly with an open upper limit expressed by `infinity`. The boundary values always belong to the range. The `length` attribute must be used in conjunction with other matching expressions, for example the 'any' matching expression or a reference to a

Table 10.16   Matching for `set of` templates

```
// example set of type
type set of integer SetOfInt;

// template examples expecting for a set of value
template SetOfInt a_oneAndAnother  := { 1, ? };
template SetOfInt a_containsOne     := { 1, * };

template SetOfInt a_oneAndAnotherB := { ?, 1 };     // same as
                                                    // a_oneAndAnother
template SetOfInt a_containsOneB   := { *, 1, * }; // same as a_containsOne

// length of 1 != 2
b := match( { 1 },        a_oneAndAnother ); // b evaluates to false,
b := match( { 2,1 },      a_oneAndAnother ); // b evaluates to true
// length of 3 != 2
b := match( { 3,2,1 },    a_oneAndAnother ); // b evaluates to false,

b := match( { 1 },        a_containsOne ); // b evaluates to true
b := match( { 2,1 },      a_containsOne ); // b evaluates to true
b := match( { 3,2,1,1 }, a_containsOne ); // b evaluates to true
```

Table 10.17   Example template definitions with `length` matching attribute

```
type record GenericParam {
  charstring id,
  charstring pValue optional
}

type set length of GenericParam GenericParam_List;

const integer c_maxPars := 2;
template GenericParam_List a_1orMaxPars := ? length ( 1 .. c_maxPars );

const GenericParam_List c_1Param := { { "tag", "abcdefg" } };
const GenericParam_List c_3Pars  := { { "branch", "1234xyz" },
                                       { "tag", "123xyz"    },
                                       { "tag1", omit       } };

b := match( c_1Param, a_1orMaxPars ); // b evaluates to true
b := match( c_3Pars , a_1orMaxPars ); // b evaluates to false
```

template definition containing 'any' or 'any-or-none'. Examples of templates with length restrictions are shown in Table 10.17.

10.4.6.3   Permutation

The `permutation` expression will match a received value list if it contains all the specified list element values, but in an arbitrary order. A `permutation` expression can

Table 10.18 Matching permutations

```
type record of integer IntSequence;

// template examples expecting a record of value
template IntSequence a_containsOneTwoUnordered  := permutation( 1, 2  );
template IntSequence a_containsOneTwoOther       := permutation( 1, 2, * );

b := match( { 1,2 },    a_containsOneTwoUnordered ); // b evaluates to true
b := match( { 2,1 },    a_containsOneTwoUnordered ); // b evaluates to true
b := match( { 1,2,2 }, a_containsOneTwoUnordered ); // b evaluates to false
// additional element
b := match( { 1,2,2 }, a_containsOneTwoOther  ); // b evaluates to true
```

be used as a template for a record of type. In a permutation expression values, templates, as well as the matching expressions 'any' and 'any-or-none' can be used.

If the permutation expression does not contain 'any-or-none', then the match is successful if there is a one-to-one mapping between the elements in the permutation expression and the record of value. If the permutation expression contains 'any-or-none', then the match is successful if there is a one-to-one mapping between the elements in the permutation expression and a subset of the record of value.

The examples in Table 10.18 are based on the record of type IntSequence introduced in Table 10.16. In the first two examples it is shown that the order in the value does not matter. But if an additional element is added, that is the template and the value are no longer of equal length and no one-to-one mapping is possible, then the match will fail. In the last example, if 'any-or-none' is used in the permutation expression, then the template and the value can be of different length because only a subset of the value is compared against the template.

Note that performing a match against a permutation expression can be computational expensive and care should be taken when using permutation expressions with a large lists of elements.

10.4.6.4 Subset and Superset

The subset expression will match a received list value if it contains at *least* one occurrence of *one* of the specified list element values. These list element values are supplied as the argument to this expression. The superset expression will match a received list value only if it contains *at least all* the specified values. Note that it will also accept additional values without problems. In TTCN-3, the subset and superset expressions may contain values and templates, except the matching expressions omit, 'any-or-none', subset and superset. Neither can the matching attributes ifpresent and length be used inside a subset or superset expression. Each of the disallowed expressions could resolve to an empty set of values, making the subset or superset expression meaningless. Note that a length restriction for the complete subset or superset expression is possible.

Both expressions may only be used in set of template definitions. The examples in Table 10.19, which are based on the set of type GenericParamList,

Table 10.19   Template definitions with `superset` and `subset` matching expressions

```
const GenericParam_List c_1Param := { { "tag", "abcdefg" } };
const GenericParam_List c_2ParsA := { { "tag", "abcdefg" },
                                       { "comp", "sigcomp" } };
const GenericParam_List c_2ParsB := { { "branch", "1234xyz" },
                                       { "tag", "abcdefg" } };


template GenericParam_List a_tagOrBranchPar :=
          subset( { "tag", "abcdefg" }, { "branch", "1234xyz" } );

b := match( c_1Param, a_tagOrBranchPar ); // b evaluates to true
// correct tag param is enough
b := match( c_2ParsA, a_tagOrBranchPar ); // b evaluates to true
b := match( c_2ParsB, a_tagOrBranchPar ); // b evaluates to true


template GenericParam_List a_atLeastSigCompPar :=
          superset( {"comp", "sigcomp"} );

b := match( c_2ParsA, a_atLeastSigCompPar ); // b evaluates to true
// since comp param missing
b := match( c_2ParsB, a_atLeastSigCompPar ); // b evaluates to false


template GenericParam_List a_atLeastSigCompParStar :=
          { { "comp", "sigcomp" }, * };
```

illustrate the differences between the two expressions. In these examples, the template `a_tagOrBranchPar` matches the value `c_2ParsA` since the matching of one list element suffices. The matching of `a_atLeastSigCompPar` fails since there is no `"comp"` parameter among the two elements in the list value `c_2ParsB`.

The cautious reader may have noticed that the `superset` expression actually has the same effect as using a 'any-or-none' matching expression. Therefore, our example template `a_atLeastSigCompParStar` matches the same values as the template `a_atLeastSigCompPar`.

10.4.7   More about Matching Expressions for String Types

Template definitions for string types allow a number of extra matching expressions. In addition, the previously introduced 'any' and 'any-or-none' matching expressions have a special meaning when applied *within* string values.

10.4.7.1   The 'any' and 'any-or-none' Matching Expressions within String Types

When the 'any' matching expression is used *within* a string template, it matches any single string character at that position. Similarly, the 'any-or-none' matching expression allows entire (sub)strings of arbitrary length to be matched within a string value. The 'any-or-none' expression also allows strings of length zero. Note that for `charstring`

Table 10.20 String template definitions with any matching expression

```
template octetstring a_2ndByteDontCare := 'AF?'O;

b := match( 'Affe'O,   a_2ndByteDontCare ); // b evaluates to true
 // 0th octet != 'AF'O
b := match( '8989'O,   a_2ndByteDontCare ); // b evaluates to false,
// 3 != 2 octets
b := match( 'Caffee'O, a_2ndByteDontCare ); // b evaluates to false,

template charstring a_A2DontCaresC := pattern "A??C";
b := match( "A->C", a_A2DontCaresC ); // b evaluates to true
// 3 != 4 letters
b := match( "AAC",  a_A2DontCaresC ); // b evaluates to false,

template charstring a_startWithAEndA := pattern "A*A";
b := match( "ABBA", a_startWithAEndA ); // b evaluates to true
b := match( "AA",   a_startWithAEndA ); // b evaluates to true

template bitstring a_startWith1End0x := '1*0?'B;
b := match( '101'B,   a_startWith1End0x ); // b evaluates to true
b := match( '11100'B, a_startWith1End0x ); // b evaluates to true
// 3rd bit != '0'B
b := match( '11111'B, a_startWith1End0x ); // b evaluates to false,
```

and universal charstring values, each occurrence of these matching expression matches not just alphanumeric characters but the full character set, that is also control characters, ':', ')' and so on.

Table 10.20 shows some example templates. The template a_2ndByteDontCare shows that for octetstring values a single character is defined as one octet, that is two hexadecimal digits. The template a_A2DontCaresC illustrates the use of these matching expressions in charstring or universal charstring. As can be seen, to use these matching expressions in this case the additional pattern keyword is required. This is necessary because '?' and '*' are also valid values for these string types. Without the pattern keyword, these characters are treated as normal letters and will therefore only match themselves.

10.4.7.2 Length Restrictions of String Values

As for list-like types, the length matching attribute can be used to further restrict the possible matches of a string template. Also as for list-like types, length has to be combined with a string matching expression. Note that care is needed when specifying the actual length restriction because the meaning of the number in the operator argument differs for each string type. For a (universal) charstring value, the unit is the number of characters, for an octetstring value, it is the number of octets, for a hexstring value, it is the number of hexadecimal digits and for bitstring, it is the number of bits. Examples of templates for string types with length matching expression can be seen in Table 10.21.

Table 10.21 Using the `length` matching attribute with strings

```
template charstring a_char7UserName := ? length( 7 );

// 6 != 7 characters
b := match( "Stefan", a_char7UserName ); // b evaluates to false,
b := match( "Stephan", a_char7UserName ); // b evaluates to true

template octetstring a_1To2BytesA := ? length( 1 .. 2 );
template hexstring   a_1To2BytesB := ? length( 2 .. 4 );

// 3 != 1..2 octets
b := match( '010203'O, a_1To2BytesA ); // b evaluates to false,
// 6 != 2..4 hex digits!
b := match( '010203'H, a_1To2BytesB ); // b evaluates to false,

templates bitstring a_byteWithPair := '*11*'B length ( 8 );

b := match( '00101011'B, a_byteWithPair ); // b evaluates to true
// 6 != 8 bits
b := match(   '001100'B, a_byteWithPair ); // b evaluates to false,
```

10.4.7.3 Escape Sequences in Text Templates

Within `pattern` expressions, TTCN-3 also supports the use of character escape sequences. These can be used to specify some pre-defined character value restrictions, unprintable or Unicode characters outside the range of 7-bit ASCII, or the literal occurrence of special characters (for example '?' or '*') in a text template. Note that escape sequences can only be used within `pattern` expressions, not within TTCN-3 text value definitions. A list of all available escape sequences is shown in Table 10.22.

Table 10.22 List of TTCN-3 character escape sequences

Sequence	Description	Sequence	Description
\?	Question mark character (?)	\\	Backslash character (\)
*	Asterisk character (*)	\", " "	Double quote character (")
\[	Open square bracket character ([)	\]	Closing square bracket character (])
\{	Open curly parentheses character ({)	\}	Closing curly parentheses character (})
\(	Open parentheses character (()	\)	Closing parentheses character ())
|	Vertical bar character (|)	\#	Hash character (#)
\r	Carriage Return character	\t	Tabulator character
\n	Line Feed character	\d	Any digit character (0 .. 9)
\w	Any alphanumeric character (0 .. 9, a .. z, A .. Z)	\q{g,p,c,r}	Unicode character with group, plane, cell and row coordinates
\b	Word boundary	\s	Any white-space character

Table 10.23 Text template definitions with character escape sequences

```
template charstring a_1AndAlphaChar := pattern "1\w";
b := match( "1a", a_1AndAlphaChar ); // b evaluates to true
// equal sign is not alphanumeric
b := match( "1=", a_1AndAlphaChar ); // b evaluates to false,

template charstring a_0thru9 := pattern "\d";
b := match( "8", a_0thru9 ); // b evaluates to true
// a is not a digit
b := match( "a", a_0thru9 ); // b evaluates to false,

template universal charstring a_stringWithSweAO := pattern
                                           "*\q{0,0,0,162}*";
b := match( "skål", a_stringWithSweAO ); // b evaluates to true
// has no å in it
b := match( "fem",  a_stringWithSweAO ); // b evaluates to false,

template charstring a_HelloQuestion := pattern "\"Hello\?\"";
b := match( """Hello?""", a_HelloQuestion ); // b evaluates to true
// "!" != "?"
b := match( """Hello!""", a_HelloQuestion ); // b evaluates to false,
```

Note that the double quote character can be escaped both by a backslash character as the other escape sequences as well as with an additional double quote character as in ordinary character strings.

A few simple example templates using these escape sequences are shown in Table 10.23. In these examples, the template a_HelloQuestion illustrates that a double quote character can be escaped differently in pattern expressions to that in a specific TTCN-3 charstring value, that is using a backslash instead of an extra double quote.

10.4.7.4   Concatenation of String and List Type Templates

When defining long or complex templates it is often useful to break the expression into a series of distinct pieces. This can be achieved in TTCN-3 by using concatenation. Template concatenation is specified by using the '&' as a separator between individual template specifications or template references. Concatenation can be used to create templates of type bitstring, octetstring, hexstring, charstring, universal charstring, record of, set of and array. The concatenation operator can also be used with pattern specifications. Examples of this are shown in Table 10.24.

10.4.7.5   Regular Expressions in Text Templates

The escape sequences together with the 'any' and 'and-or-none' matching expressions seen so far can be considered as a simple form of regular expressions. TTCN-3, however, provides a more expressive form of regular expressions for text string templates, that is based on the already-introduced pattern construct. The use of these regular expressions

Table 10.24 Template definition with concatenation

```
template charstring a_start := "start ";
template charstring a_now := "now";
template charstring a_startnow := a_start & a_now & "?";


template charstring a_alphaChar := pattern "\w";
template charstring a_digitChar := pattern "\d";

template charstring a_complexPattern := pattern a_startnow & a_alphaChar &
                                        a_digitChar;

b := match( "start now?A1", a_complexPattern ); // b evaluates to true
```

is restricted to TTCN-3 charstring and universal charstring values only, they cannot be used for the binary string types. The operators in Table 10.25 allow complex regular expressions to be constructed, starting from simple expressions or strings.

Related to regular expressions is the regexp operation. This operation offers the possibility of extracting substrings from a given text value or a template that specifies a specific value. The arguments to this operation are the text value to be evaluated, a pattern expression, that is to be matched, and the index of the matching substring instance within the text value that should be returned. Our example in Table 10.26 specifies a template with a regular expression that matches any string where "Name: " is followed by either a tabulator or a space, and then at least one alphabetic character. The invocation of the regexp operation then extracts the part of the string value, that is the first match from the beginning for the expression within the template definition, that is the alphabetic string "Stephan". The use of incorrect arguments to this function, for example a negative substring index or a template containing wildcards, leads to a test case error. Indexing of the matched parts starts from 0.

10.5 Template Definitions for Signatures

Templates for sending and receiving values in procedure-based communication, that is signature templates, are essentially specified in the same way as their message counterparts.

Signature templates used in the sending operations call and reply must specify a specific value for all outgoing parameters, that is in and inout parameters in a call operation or out and inout parameters in a reply operation. Contrary to message templates, however, such signature templates must also address parameters that pass information in the opposite direction and are not relevant in these calls. The values of these kinds of parameters should be specified using the unspecified value '–'.

Signature templates used in receiving operations, such as getcall and getreply, may either use a specific value or a matching expression for all incoming parameters. Parameters in the opposite direction that are not relevant to the calls can be specified using either an arbitrary matching expression or again the unspecified value '–'. Note that these parameters are simply ignored by the matching mechanism.

Table 10.25 Text template definitions with regular expressions

Expression	Description	Examples
[char1char2]	Set of characters, "-" is used for character ranges, "^" for exclusion of characters	**template charstring** a_AorB := **pattern** "[AB]"; **template charstring** a_lowerChar := **pattern** "[a-z]"; **template charstring** a_noLowerChar := **pattern** "[^a-z]"; *// b evaluates to true* b := **match**("A", a_AorB);

(expr|expr)	Grouping and choice	**template charstring** a_digits0thru19 :=
		 pattern '(\d|(1\d))';
		
		 // b evaluates to true
		 b := **match**("13", a_digits0thru19);
		 // b evaluates to false
		 b := **match**("20", a_digits0thru19);

expr#(n, m)	Repetition expression where lower or upper boundary may be omitted if infinite	**template charstring** a_2plusDigits :=
		 pattern " \d#(2,)";
		
		 // b evaluates to true
		 b := **match**("2004", a_2plusDigits);
		 // b evaluates to false
		 b := **match**("5:00", a_2plusDigits);

expr#n	Repetition expression, exactly n repetitions	*// Same as expr#(n)* *// for shorthand expr#n n has to be* * single digit* **template charstring** a_2Digits := **pattern** " d#2";
expr+	At least one occurrence of expression	*// Same as expr#(1,)*

[range]	Character set expressions which can specify ranges (-) and/or complement (^)	**template charstring** a_3capsOrMore :=
		 pattern "[A-Z]#(3,)";
		
		 // b evaluates to true
		 b := **match**("TTCN", a_3capsOrMore);
		 // b evaluates to false
		 b := **match**(":o)", a_3capsOrMore);

(continued overleaf)

Table 10.25 *(continued)*

Expression	Description	Examples
{*ref*}	Reference expression to include constants, variable values or other template definitions	```template charstring a_language := 'TTCN'y;``` ```template charstring a_statement := pattern " {a_language} is easy!";``` ```// b evaluates to true``` ```b := match("TTCN is easy!", a_statement);``` ```// b evaluates to false``` ```b := match("Finnish is easy!", a_statement);```
\N{*ref*}	Reference a character set via a template, constant, value or subtype of (universal) charstring	```template charstring a_vowel := pattern "[aeiou]";``` ```template charstring a_twovowel := pattern := "\N{a_vowel}\N{a_vowel}"``` ```// b evaluates to true``` ```b := match("ab", a_twovowel);```

Table 10.26 Text string extraction using the TTCN-3 `regexp` operation

```
// the definition of the matching criteria
template charstring a_nameEntry :=
pattern "Name:[ \t]([A-Z][a-z]#(1,))";

// the extraction of the name yields v_name == "Stephan"
var charstring v_name := regexp( "Name: Stephan", a_nameEntry, 0 );
```

Table 10.27 Example signature template definitions

```
void update( in string key, inout string val )
raises ( NotAllowed, SessionExpired );

template update a_updateJohnsPassword := {
  key := "password of John",
  val := "pa$$w0rd"
}

template update a_anyUpdateReply := {
  key := -, // possible since 'in' parameter
  val := ?
}
```

The example signature templates presented in Table 10.27 are based on the example presented earlier in Table 6.10. The template `a_updateJohnsPassword` has been used in a `call` operation to a directory server, whereas the template `a_anyUpdateReply` checks if the directory server reply contains any value.

Return values for a procedure call as well as exceptions are not specified within a signature template. Instead, these parts of a signature require a separate template definition, which follows the form previously discussed for message templates.

10.6 Assignment, Access of Templates and the Pre-Defined Functions Isvalue and Valueof

Templates and values are different kinds of objects in TTCN-3 and are treated differently by the language. Templates and values cannot be used interchangeably: a value always specifies *exactly one* of the values allowed by its underlying type definition, whereas a template specifies a *subset of all* values allowed by the underlying type definition (this subset may contain as many as all values or as little as a single value). In short, a value can be used in those places where a template is required (in which case it is implicitly turned into a template), but it is not possible to use templates in places where values are required. As an example, a template cannot be used as an actual parameter of a function or another template, if the formal parameter has not been declared as a template parameter. Similarly, templates cannot be used in assignments to variables, conditions or expressions. It is also not possible to read or change the fields, elements or alternatives in a template using dot or index notation. It is thus *not* allowed to write:

```
var charstring v_hostName := a_nokiaAt5060.host.hostName;
```

A template can be turned into a value using the pre-defined function `valueof`, given that the template only specifies a single value for all fields. The following example, shown in Table 10.28, is therefore valid since the template `a_nokiaAt5060` specifies only a single value. Should the template contain any matching or regular expressions, the execution of `valueof` will cause a test case error. Such errors can be prevented by first checking the template to be converted with the pre-defined function `isvalue` which returns either `true` or `false`. This function is also useful in the context of other operations which require specific values such as `send` operations. For example, it can be used to ensure that complex parameterised template definitions, which will be introduced in more detail in our next chapter, fulfil at run-time this constraint.

Table 10.28   Using `valueof` and `isvalue`

```
var HostPort v_currentHostPort := valueof( a_nokiaAt5060 ); // OK
var HostPort v_hostAndPortErr  := valueof( a_nokiaAtAnyPort ); // ERROR

if ( isvalue(a_nokiaAtAnyPort) ) {
  // this log statement should never be printed!
  log("a_nokiaAtAnyPort specifies a specific value!");
}
```

Table 10.29 Types and applicable matching expressions

	Specific value	Value list/ complement	Value range	any ('?')	any-or-none ('*')	omit/ ifpresent	length	pattern	subset/ superset
integer, float	●	●	●	●		●[b]			
(universal) charstring	●	●		●		●[b]	●	●	
bitstring, hexstring, octetstring	●	●		●		●[b]	●		
within bitstring and so on	●			●	●				
record, set, union	●	●		●		●[b]			
within record, set	●	●	(●)	●	●	●	(●)	(●)	
within union	●	●	(●)	●			(●)	(●)	
record of, set of	●	●		●		●[b]	●		●[a]
within record of, set of	●	●	(●)	●	●	●[b]	(●)	(●)	
enumerated, boolean, anytype, verdicttype	●	●		●		●[b]			

[a] Subset and superset matching expressions are only possible on set of types.
[b] Omit can only be assigned to template variables of this type which are introduced in the next chapter.
Note: For entries marked with "(●)" the use of the matching expression is only allowed where the corresponding field or element type permits it.

Template restrictions in TTCN-3 will ease the analysis of TTCN-3 code for such errors. These are considered as a more advanced topic and are explained in more detail in Section 11.6.

10.7 Summary

In this chapter, we have provided the basics for the specification of templates as well as their handling by the TTCN-3 matching mechanism. We have firstly shown that templates can specify a specific message, procedure call, reply and so on. Secondly, we have shown that templates can be specified to allow multiple values for one or more parts of a given message. We have seen that the latter kind of templates use so-called matching expressions and can only be used in receiving operations.

Matching expressions are a powerful feature that distinguishes TTCN-3 from other programming languages and allows the test engineer to focus only on the relevant parts of a message when receiving data from the SUT. TTCN-3 offers a variety of matching expressions that in some cases are restricted in their use to specific underlying types. Among the covered expressions are value lists, value ranges, 'any' and 'any-or-none'. Also, we introduced regular expressions for text string matching that are a new addition to the TTCN language. Finally, we clarified the difference between templates and values.

Because of the abundance of different matching expressions and the size of the TTCN-3 type system, we close this chapter with Table 10.29, which attempts to summarise the relationship between the two. This table may also be useful as a quick reference guide when implementing templates.

11

Advanced Templates

In practice, information content, that is communicated between the test components and the system under test (SUT) is rarely so simple that it can be expressed using simple unstructured types. The more common case is that such content is specified by deeply nested type structures, which require the specification of quite complex templates. In these cases, templates tend to grow large and the use of one template definition per message quickly becomes hard to read and maintain.

To manage complex values in a better way, TTCN-3 offers the possibility to parameterise template definitions and to decompose templates using template references. Furthermore, it is possible to define templates via the selective modification of existing templates. This provides yet another powerful way for a more concise definition of message families. Finally, templates can even be computed at runtime by template computing functions. Template restrictions increase the possibilities to analyse templates, especially whether templates in `send` statements correspond to single values only. After introducing these mechanisms in this chapter, we will provide some guidance on how to use them to structure templates for complex messages and discuss the issue of implicit template definitions.

11.1 Template Definitions for Complex Type Structures

So far in this book, we have mainly focused on template definitions for fairly simple type structures, with at most a handful of fields. In practice, however, we commonly encounter message structures that have many levels of user-defined types. Templates for such types can be defined by nesting values and matching expressions in the way defined by the underlying type structure. The beginning and end of each individual structured type has to be indicated using curly brackets as shown in Table 11.1.

The example shown in Table 11.1 brings us a step closer to what template definitions may look like in the real world. But this template still only specifies a fraction of a complete Session Initiation Protocol (SIP) message, namely, a SIP Uniform Resource Identifier

An Introduction to TTCN-3, Second Edition.
Colin Willcock, Thomas Deiß, Stephan Tobies, Stefan Keil, Federico Engler and Stephan Schulz.
© 2011 John Wiley & Sons, Ltd. Published 2011 by John Wiley & Sons, Ltd.

Table 11.1   Verbose value template for a complex type

```
type record SipUri {
  UserInfo            userInfo    optional,
  HostPort            hostPort,
  GenericParam_List urlParams   optional,
  GenericParam_List urlHeaders optional
}

type record UserInfo { charstring user, charstring passwd optional }

type record HostPort { charstring host, integer portNumber optional }

union IpAddress { charstring ipv4, charstring ipv6, charstring hostName }

type set length ( 1 .. infinity ) of GenericParam GenericParam_List;

type record GenericParam { charstring id, charstring pValue optional }

template SipUri a_stephanSipUriAtAnyHostWithSigComp := {
  userInfo  := {
    user    := "stephan",
    passwd := omit
  },
  hostPort := {
    host         := { hostName := ? },
    portNumber := 5060
  },
  urlParams   := {
    [0] := {
       id      := "comp",
       pValue := "sigcomp"
    }
  },
  urlHeaders := omit
}
```

(URI). Therefore, a template definition for a complete SIP message will be even bigger in terms of complexity and size. One way to reduce the size of this specific template would be to use the alternate list format for specifying the values and matching expressions:

```
template SipUri a_stephanSipUriAtAnyHostWithSigComp := {
    {"stephan", omit},
    { { hostName := ? }, 5060 },
    { { "comp", "sigcomp" } },
    omit
}
```

However, this approach will at best only reduce, but not solve, our problem with the specification of the complete SIP message. In addition, this alternate list format raises questions about readability and future maintainability of such a template specification.

Another issue we face is that it is often necessary to have many variations on such a template definition, for instance, when the same fields are used in several different parts of a message. SIP URIs, for example are used in many places within SIP messages. In addition, similar templates are often needed for similar message exchanges in different test cases. Therefore, instead of just using the compact list syntax, we advocate the use of the advanced template definition mechanisms that TTCN-3 offers and which will be introduced in the following sections.

11.2 Template References

Arguably the most commonly used advanced template definition mechanism is the use of template references. For complex types, a single template definition may easily cover multiple computer screens and become hard to read, as well as to maintain. In addition, template definitions for complete messages tend to be only usable for one specific communication operation. Typically, they can neither be reused in other communication operations within the same test case nor be used across multiple test cases. However, parts (like certain fields) of such templates may be recurring in different situations and should be reused wherever possible.

Template references allow decomposing a single template definition into multiple smaller template definitions. Other templates can be referenced from a given template by simply replacing the specification of a field (or element) value by the identifier of another template. Table 11.2 shows the complex template from Section 11.1 decomposed using

Table 11.2 Decomposition of a template definition with template references

```
template SipUri a_stephanSipUriAtAnyHostWithSigComp := {
  userInfo   := a_stephanUser,
  hostPort   := a_anyHostAtPort5060,
  urlParams  := a_sigCompParam,
  urlHeaders := omit
}

template UserInfo a_stephanUser := {
  user   := "stephan",
  passwd := omit
}

template HostPort a_anyHostAtPort5060 := {
  host       := { hostName := ? },
  portNumber := c_sipPort
}

template GenericParam_List a_sigCompParam := {
    { id := "comp", pValue := "sigcomp" }
}
```

template references. These four template definitions are not just easier to read. The templates, which are referenced from the `a_stephanAtAnyHostSipUriWithSigComp` template, can now also be referenced for other purposes from other template definitions.

At every point within the TTCN-3 code where it is possible to specify a direct template reference, it is also possible to specify a call to a function that returns a template.

When the structure of a message type is complex, the selection of the granularity for template definitions is not a trivial design decision. The decomposition of a single message into too many template definitions may negatively affect readability. At the same time, the reluctance to break up templates into relevant, smaller definitions may also hinder their reuse. We will give some guidance on this issue in the last section of this chapter.

11.3   Template Parameterisation

It is possible to parameterise templates in a similar way to functions. Parameters can be used to pass regular values, other templates and matching expressions into a template definition. Parameters specify information that only becomes definite during test system execution, that is the parts of a template definition that use this information are only fixed once an operation uses the template.

Parameters of templates are always `in` parameters and cannot be of `out` or `inout` kind. The `in` keyword may be used in the template parameter definition to indicate this fact. Within a parameterised template definition, the parameter identifiers can be either directly assigned to fields or used in operations that are called inside the template definition.

11.3.1   Value Parameters

Parameters for templates can be further classified into value parameters and template parameters. Value parameters must be instantiated with proper values. Using template references or matching expressions is not allowed. Note that the `omit` value by itself is not considered to be a TTCN-3 value. The examples in Table 11.3 show the definition of templates with value parameters. It also shows how parameterised templates can be referenced from other templates or operations.

11.3.2   Template Parameters

Template parameters are defined by preceding the parameter declaration with the `template` keyword. In this case it is possible to pass in other templates, matching expressions or `omit`. Also normal values can be passed in, which are then interpreted as inline templates. On the other hand, when using template parameters, it is no longer possible to perform operations with the parameter value like arithmetic expressions or string concatenation, which has been used in the definition of the template `a_dotComHostPort` in Table 11.3.

In Table 11.4 the template `a_hostNamePort` has template parameters and thus allows instantiation with other templates or matching expressions.

Table 11.3 Template definitions with value parameters

```
template HostPort a_dotComHostPort( in charstring p_hostName,
                                    in integer    p_portNumber ) := {
  host       := p_hostName & ".com",
  portNumber := p_portNumber
}

template SipUri a_stephanSipUriAt5060WithSigComp (in charstring
                                                  p_hostName ):= {
  userInfo   := a_stephanUser,
  hostPort   := a_dotComHostPort( p_hostName, 5060 ),
  urlParams  := a_sigCompParam,
  urlHeaders := omit
}
```

Table 11.4 Template definition with template parameters

```
template HostPort a_hostNamePort( template charstring p_hostName,
                                  template integer    p_port ) := {
  host       := { hostName:= p_hostName },
  portNumber := p_port
}

template charstring a_anyHostName := pattern "[\w.]#(1,)";

template SipUri a_stephanSipUriAtAnyHostNameWithSigComp := {
  userInfo   := a_stephanUser,
  hostPort   := a_hostNamePort ( a_anyHostName, * ),
  urlParams  := a_sigCompParam,
  urlHeaders := omit
}
```

11.3.3 About the Use of Template Parameterisation

Similar to the case of template references, the selection of what parts of a template should be parameterised and in which manner is a non-trivial design decision. The combination of decomposing a complex message into multiple template definitions with associated parameterisation can be critical in achieving broad reuse of repeatedly used message structures. Overusing parameterisation may, however, have a negative impact on the readability of TTCN-3 code.

11.4 Selective Modification of Other Templates

This third template mechanism helps to simplify the definition of two or more templates with mostly the same information content. It is only applicable to templates of structured or list types. A modified template inherits most of its definition from its so-called base

Table 11.5   Base template and modified template definitions

```
template GenericParam a_sigCompParam := {
  id      := "comp",
  pValue := "sigcomp"
}


template GenericParam a_anyCompParam modifies a_sigCompParam := {
  pValue := ?
}
```

template and only specifies the way in which it differs from the original. A modified template definition must reference the base template identifier after the `modifies` keyword. In Table 11.5 the definition of `a_anyCompParam` modifies the definition of the base template `a_sigCompParam`. The modification means that in the new template any value for the field `pValue` will be matched, instead of the fixed value `"sigcomp"` in the case of the original `a_sigCompParam`.

A modified template can itself be modified again and in this case it becomes the base template for this newly defined template. Cyclic dependencies of modified templates are not allowed in TTCN-3.

When modifying parameterised templates, some extra requirements are imposed on the modifying definition. The modifying template must preserve all base template parameters with their types as well as their declaration order, even if the modifying template definition no longer makes use of such parameters. This requirement is shown by the template `a_tagNoOrAnyPar` in Table 11.6. In contrast, the template `a_anyIdPar`, in the same table, violates this requirement. A parameter of the base template may have a default value or template. The modified template may use either the same default value or template or the default value or template is removed. The first case is written by simply using a dash instead of the default value or template, in the second case the assignment of a default value or template is dropped completely.

It is also possible to add additional parameters in the modifying definition. The new parameters have to be concatenated at the end of the base template parameter list as the definition of the template `a_genericPar` illustrates in Table 11.6.

Modified templates have a great potential for reducing the duplication of information between similar template definitions. On the other hand, it must be kept in mind that any future changes to the values in the base template may automatically affect a number of other template definitions (since they may be derived from the changed template via a chain of modifications). To avoid unpleasant surprises, we encourage TTCN-3 writers to clearly identify base template definitions, that is templates that are supposed to be modified by derived template definitions. This can be done by using some form of naming scheme. Secondly, templates should be selected as base templates if they have values that are not expected to change much over the lifetime of the test system. It is possible, however strongly discouraged, to use a modified template as a base template for further modifications. The longer a chain of modification becomes, the harder it will be to discern the *actual* value of a derived template.

Table 11.6 Base template and modified template definitions with parameters

```
template GenericParam a_tagPar ( charstring p_pValue ) := {
  id      := "tag",
  pValue := p_pValue
}


template GenericPar a_tagNoOrAnyPar( charstring p_pValue )
modifies a_tagParam := {
  pValue := *
}


template GenericPar a_anyIdPar( charstring p_id )
modifies a_tagParam := {
  id      := ?
} // ERROR: parameter list of modified template is missing
  // p_pValue parameter

template GenericPar a_genericPar( charstring p_pValue,
                                  charstring p_id )
modifies a_tagParam := {
  id      := p_Id,
  pValue := p_pValue
}
```

11.5   Explicit versus Implicit Template Definitions

In any place where a template is required as an argument, for example in communication operations or the instantiation of a template parameter, ordinary values can be passed instead of template identifiers. In communication operations, this means that the variable, constant or pure value in TTCN-3 value notation can specify a complete message. Such values are called 'implicit' or 'inline' templates as they more or less represent a single value template without a template identifier. Two implicit templates are shown in the example in Table 11.7. The second example shows that even matching expressions can be used in implicit templates.

We discourage the use of such template definitions in communication operations. Although they may seem to offer at times a quick and easy solution, they will in the long run, negatively affect TTCN-3 code readability and test suite maintainability. The most problematic issue is that implicit templates cannot be reused in a test suite because of the lack of a template identifier. Note that in this book we have used implicit templates to keep examples more compact, but this is not advisable in actual test suites.

Table 11.7 Send operation with implicit DNS message template

```
pt.send( DnsMessage: { 0, e_question, "www.research.nokia.com", omit } );
pt.receive( DnsMessage: { ?, e_answer, "www.research.nokia.com", * } )
```

11.6   Restricting Template Usage

Templates in general define a set of values; for templates used in receiving operations
this is the perfect notion. But for a template used in a sending operation this does not fit
exactly: A template used in a sending operation has to define exactly one value. This is
a potential cause of runtime errors in a TTCN-3 test system.

In the example in Table 11.8 there are two send operations, both using the parameterised template a_tagPar. In the first case, a single charstring parameter is used
as actual parameter and the resulting template represents a single value. In the second
case, the template a_charstringAny, representing several values, is used as actual
parameter. The resulting template will itself represent several values, which in turn will
cause a runtime error in the send operation.

A TTCN-3 tool could warn the user of this problem already before executing the code.
But when analyzing the TTCN-3 code a tool cannot know which of the templates will
be used in receiving and which ones will be used in sending operations. A tool could
perform a sophisticated data flow analysis, determining for many templates how they are
used and issue corresponding warnings. But such an analysis cannot be complete, there
will always be the possibility that a tool either misses some templates representing several

Table 11.8   Template causing a runtime error

```
template charstring a_charstringAny := ?;
template GenericParam a_tagPar ( template charstring p_pValue ) := {
   id      := "tag",
   pValue := p_pValue
}

pt.send( a_tagPar( "ok" ));
pt.send( a_tagPar( a_charstringAny ));
```

Table 11.9   Templates with restrictions

```
template charstring a_charstringAny := ?;
template (value) GenericParam a_tagPar (
   template (value) charstring p_pValue ) := {
   id      := "tag",
   pValue := p_pValue
}
template (value) GenericParam a_tagParMany (
   template (present) charstring p_aValue ) := {
   id      := "tag",
   pValue := p_pValue
}

pt.send( a_tagPar( "ok" ));
pt.send( a_tagPar( a_charstringAny ));
```

values and potentially causing a runtime error or a tool would issue warnings even for templates that will not cause problems.

To solve this problem, TTCN-3 allows there to be an indication of whether a template should represent just a single value or whether it can represent several values. Actually there are three ways to indicate such a restriction of a template: It can be restricted to a single value (`value`), it can be restricted to a single value or to `omit` (`omit`), and it can be restricted to a set of values without matching `omit` (`present`).

In Table 11.9 template restrictions have been added to the example from Table 11.7. The template `a_charstringAny` is still a template representing several values, no template restriction has been added. The template `a_tagPar` has been marked as a template representing a single value only, the restriction (`value`) has been added to its definition. Correspondingly the parameter of this template has also been marked as a template representing a single value. In the template `a_tagParMany` this parameter has been restricted such that `omit` cannot be passed as parameter by the restriction (`present`), but any other template even representing several values would be accepted. As such it cannot be guaranteed that `a_tagParMany` is a single value and TTCN-3 tool could safely raise a corresponding warning. The second `send` statement would cause a similar warning because the actual parameter used – `a_charStringAny` – is a template representing several values, whereas the formal parameter `p_aValue` was restricted to be a template representing a single value only.

Note that a TTCN-3 tool is not required to perform such checks at compile time, although the template restrictions have been introduced for exactly this purpose: Allowing a more thorough analysis of the TTCN-3 code before executing it. Note also that template restrictions are an optional mechanism, authors of TTCN-3 code are not obliged to use them.

11.7 Template Variables and Computing Functions

The most advanced and dynamic possibility to define templates in TTCN-3 is to actually modify and compute templates at runtime. Functions in TTCN-3 can also be used to compute templates. Inside a function it is possible to declare template variables. Templates can also be modified at runtime and returned as a result of a function.

Table 11.10 Computing a template

```
function f_tagPar ( charstring p_pValue )
  return template GenericParam {
  var template GenericParam v_par := ?;
  v_par.id      := "tag",
  v_par.pValue := p_pValue;
  return v_par;
}

function f_idOnly ( inout template GenericParam p_par ) {
  var template charstring v_pValue := omit; // OK for template variable
  p_par.pValue := v_pValue;
}
```

Table 11.11   Implicitly setting template fields

```
function f_tagParAnyValue ( charstring p_pValue )
  return template GenericParam {
  var template GenericParam v_par := *;
  v_par.pValue      := p_pValue,
  return v_par;
}
template GenericParam a_tagParMany ( charstring p_pValue ) := {
  id      := ?,
  pValue := p_pValue}
```

In the example in Table 11.10 the function f_tagPar returns a template of type GenericParam. Within the function at first the template variable v_par is declared and initialised to match any values, thereafter the two fields are set to specific values, and the computed template is returned. The function f_idOnly shows that the special symbol omit may be assigned to template variables of any type. Such template variables can however then only be used with optional fields like pValue in this example or will otherwise cause a test case error.

When assigning templates to previously undefined parts, some other parts might become implicitly defined. The function f_tagParAnyValue in the example in Table 11.11 is a slight variation of the previous example. The template variable v_par is initialised to 'any-or-none' instead of 'any'. Before assigning the actual parameter to the field pValue, the template variable v_par could also hold omit. After this assignment this is no longer possible, as this field does have a value. At this point the field v_par.id is implicitly defined, as the field is a mandatory one it is defined to 'any'. Therefore the result of f_tagParAnyValue is the same as the template a_tagParMany.

Template restrictions, see Section 11.6, can be used together with template computing functions, but have not been used here for the sake of simplicity. In many cases, the use of template computing functions can be avoided by performing the computation separately as value computing functions and passing the result via parameters into a template. Therefore template computing functions are seldomly needed.

11.8   Structuring of Template Definitions for Complex Types

As mentioned in the previous sections, the decomposition and parameterisation of templates has a crucial impact on the readability and reusability of your TTCN-3 code. Getting this right will be challenging for newcomers to TTCN-3. Our experience is that in the real world message structures are typically more complex than those shown in our examples so far. Additionally, real test suites have hundreds or even thousands of TTCN-3 test cases that can contain a substantial number of message exchanges.

In general, a simple guideline to follow is to decompose large template definitions only once the same information starts to be repeated across multiple templates. Once this repetition is identified, this part of the template is a candidate to be factored into a separate definition. Messages that are recurring with only minor variations are typical candidates to be turned into parameterised templates. In practice, it is good to avoid the two extremes of specifying one parameterised template per type definition or using one single template definition for each complete message definition.

One possibility in protocol testing is to base the decomposition on major information elements identified by the protocol standard, for example headers in the case of SIP. In such an approach, there would be template definitions completely defining each information element. Some aspects of the information in these templates could be parameterised to increase their potential reuse. A protocol message could then be specified in a separate template definition, which uses template parameters to pass its information elements into the message. Such added flexibility in the message template definition can enable test case writers to use this template in multiple communication operations. Section 15.3 further elaborates on this topic of writing readable and reusable template definitions. This subject is discussed in the context of the SIP protocol, which has relatively complex message type definitions.

Another issue is the naming of template definitions. We suggest the use of a naming convention to identify templates that may only be used in receiving communication operations. We also suggest the identification of template definitions that are used as base templates by other template definitions. These simple measures will help you, as well as other people, to locate mistakes in test case specifications as well as to reuse existing template definitions in the development of templates for new test cases in the test suite. Such naming conventions can be combined well with template restrictions as presented in Section 11.6.

11.9   Summary

This chapter introduced the concepts of template referencing, template parameterisation and modification of templates. Modification of templates is possible by explicitly defining a template as a modified version of another, but also by changing its content in an assignment or by computing it dynamically. Template restrictions provide a way to analyse TTCN-3 code more precisely and thereby avoiding errors detected only at runtime. These advanced template mechanisms are very useful in reuse and management of TTCN-3 code in large test suites as well as test suites, in which messages with large information content are exchanged. The best use of these concepts in the specification of a test suite is application dependent. We provided some guidelines on how templates could be decomposed and parameterised.

12

Extension Packages

TTCN-3 has been extended several times since its initial definition. Some of the extensions are clearly separated from the core language and have been defined in new parts of the standard. Examples of such extensions are the IDL to TTCN-3 mapping [14] and the documentation comments [16]. Other extensions are closely related to already existing concepts of the core language and have been included in the definition of the TTCN-3 core notation. The definition of template variables and the definition of alive test components are examples of such extensions. Lately, TTCN-3 has been extended by several new concepts, such as behaviour types and support for real-time testing. These concepts have been defined in separate *extension packages* to keep the core notation stable and to avoid the situation where all TTCN-3 users would have to care about them, even when not needing them.

Each extension package defines the new concepts as additions to the core notation. But it also provides the corresponding additions to other parts of the standards: Corresponding to the definition of the concepts in the core notation their semantics have to be defined in the operational semantics. In most cases, the extensions also require additions to the TRI TTCN-3 Runtime Interface and TCI TTCN-3 Control Interface, for example by defining the corresponding logging operations for the newly defined concepts. As such, each extension package is a self-contained document, extending several parts of the standard.

At the time of writing this book four extension packages have been defined.

- **Configuration and deployment support** [19], Section 12.1: This extension package introduces configurations of parallel test components which keep their state across the execution of several test cases. Executing test case after test case on the same test configuration, building upon the outcome of previous test cases, is suitable for a more explorative or interactive kind of testing.
- **Performance and Real Time Testing** [21], Section 12.2: A notion of global time and operators to access the current time is introduced by this extension package. It becomes possible to suspend operation of a parallel test component until a specific time is reached, several parallel test components can proceed with their operation at

An Introduction to TTCN-3, Second Edition.
Colin Willcock, Thomas Deiß, Stephan Tobies, Stefan Keil, Federico Engler and Stephan Schulz.
© 2011 John Wiley & Sons, Ltd. Published 2011 by John Wiley & Sons, Ltd.

the same time. Additionally, the time from sending a message to receiving a reply can be measured more precisely.

- **Advanced Parameterisation** [18], Section 12.3: This extension package defines two aspects of type parameterisation. Firstly, it allows a type definition to be parameterised by value parameters. For example, the length of a record of type could be provided by a parameter in the type definition. Secondly, types can be used as parameters in other type definitions as well as in other definitions in general. This allows libraries of TTCN-3 code to be written, where container data types and operations on them are defined without specifying the data actually contained.
- **Behaviour Types** [17], Section 12.4: This extension package extends the type system with types for behaviours, that is functions, altsteps and test cases. Behaviour types can be used to define TTCN-3 libraries, where certain aspects of the behaviour are provided only when instantiating the library. Callback functions are a typical example, which are called from within the library, but defined outside of it.

It is possible to use just a single extension package, but it is also possible to use several extension packages in combination with each other.

12.1 Static Test Configurations

So far we have seen that the configuration of a TTCN-3 test case, which is the test components, their mappings and connections, is created dynamically by explicit operations. This fits well to a testing paradigm where test cases are considered as self-contained and the execution of a test case does not depend on the previous execution of other test cases. But sometimes this self-containedness is too strict. For example, when a user wants to execute test cases interactively it would be more convenient to start a couple of test components and execute the test cases on this already started test components without having to terminate and to restart the test components again and again. This would give the user more freedom in choosing the sequence of test cases interactively without having to define a new test case.

Static Test Configurations [19] allow a configuration of test components to be defined with mappings and connections, to create and destroy such a configuration, and to execute several test cases sequentially on such a configuration. The state of the test components is persistent, that is component variables, timers and so on, are not reset at the start of a new test case. Only verdicts and the lists of activated defaults are reset.

A static test configuration is defined by the function, that is used to create it. Table 12.1 defines a static test configuration corresponding to the example shown in Chapter 5. The function is of a specific type configuration. Similar to a test case definition it has to be specified what is the type of the main test component and of the test system interface. The main test component in this example is of type StaticComponent, which is defining just three component variables to hold references to parallel test components. These parallel test components are created as static test components when executing the configuration function, and also the mappings among ports of the static test components and the test system interface are created as static mappings. This is expressed by the keyword static following the corresponding operations.

Table 12.1 Static test configuration

```
type component StaticComponent {
    var DNSEntity v_client;
    var DNSEntity v_root;
    var DNSEntity v_remote;
}
configuration stc_threeComponents () runs on StaticComponent
system ConcurrentDNSTester {
    // create all parallel test components
    v_client := DNSEntity.create static;
    v_root   := DNSEntity.create static;
    v_remote := DNSEntity.create static;

    // map the ports of the ptcs to the test system interface
    map( v_client:pt, system:pt_client ) static;
    map( v_root:pt,   system:pt_root ) static;
    map( v_remote:pt, system:pt_remote ) static;
}
```

Static test components, mappings and connections can be created only within configuration functions. A test case executing on a static test configuration may still create further test components dynamically and it may create mappings and connections among both statically and dynamically created test components. But these dynamically created test components will be terminated at the end of the test cases, whereas the static test components will continue to exist. Similarly, the dynamically created mappings and connections are undone at the end of the test case.

Note that a configuration function may have parameters and the creation of static test components may depend on the parameter values. As such, static test configurations are less static than their name might imply. Nevertheless, once created, a static test configuration does not change throughout its lifetime.

To create a static test configuration, the corresponding configuration function is executed in the control part of a TTCN-3 module. If the function can be executed successfully it returns a handle to the configuration. This handle can be used to execute test cases and to destroy the test configuration. This is shown in Table 12.2, a test configuration is created by the function `stc_threeComponents`, the handle is assigned to a variable, and later on the stored handle is used to destroy the test configuration.

Table 12.2 Creating a static test configuration

```
control {
    var configuration dnsConfiguration := null;
    dnsConfiguration := stc_threeComponents();
    // execute test cases
    dnsConfiguration.kill
}
```

Table 12.3   Test case on static test configuration

```
testcase tc_locallyUnresolvedStatic () execute on stc_threeComponents
{

    const Question        c_clientQuestion := "www.research.nokia.com";
    const Answer          c_clientAnswer   := "172.21.56.98";
    const Question        c_rootQuestion   := "ns.nokia.com";
    const Answer          c_rootAnswer     := "131.228.6.229";
    const Identification c_identification := 12345;


    timer t_guard;

    // start the behaviour on the parallel test components
    v_root.start  ( f_server( c_rootQuestion,   c_rootAnswer ) );
    v_remote.start( f_server( c_clientQuestion, c_clientAnswer ) );
    v_client.start( f_client( c_clientQuestion, c_clientAnswer,
                              c_identification ) );

    // wait until all parallel test components are done, at most 30 seconds
    t_guard.start( 30.0 );

    alt {
      [] all component.done {
           t_guard.stop;
           // use verdicts of parallel test components
         };
      [] t_guard.timeout {
           all component.stop;
           setverdict( fail );
         }
      };
}
```

Test cases can be executed on static test configurations. The test configuration is provided by an `execute` on clause in the test case definition. The body of the test case is executed on the main test component as defined in the configuration function. Similarly the test system interface of the configuration function is used as a test system interface of the test case. In Table 12.3 the parallel test components have been created already when creating the test configuration. The parallel test components can be referred to via the component variables v_root, v_remote and v_client of the main test component and used to start test behaviour on the parallel test components.

The test case can be executed on the static test configuration by calling it in the control part and providing the handle to the static test configuration as argument, see Table 12.4.

The specific test case in this example does not modify any of the component variables. But if it would do so, then the modification would be visible to the next test case executed on this configuration. Component variables within a static test configuration can be used to carry information from test case to test case. Note that the verdict of static test components is set to **none** at the start of a test case. Also the list of activated defaults is reset when starting a test case. All other component variables, timers and ports keep their state. Note

Table 12.4   Executing a test case on a static test component

```
control {
    var configuration dnsConfiguration := null;
    dnsConfiguration := stc_threeComponents();
    execute( tc_locallyUnresolvedStatic(), dnsConfiguration );
    dnsConfiguration.kill
}
```

also that the actual test behaviour has been described in exactly the same way in both cases. The main difference between normal and static test configurations is how and when test configurations are created and destroyed.

12.2   Real-Time in TTCN-3

Timers in TTCN-3, see Section 4.6, provide already a notion of time within TTCN-3. These timers have been sufficient to express the timing properties that occur when testing communication systems. But when testing systems with stricter real-time properties these timers are not sufficient. As an example, when measuring the time between sending a message and receiving a reply with a timer the internal processing delays will also be measured. The extension package Performance and Real Time Testing [21] additionally provides a notion of real-time to TTCN-3.

Real-time in TTCN-3 is expressed via a system wide clock. This clock is reset at the beginning of each test case and it can be accessed in each test component. The system clock and the timers as defined previously are synchronised to each other, they proceed together.

The necessary precision of the clock can be expressed for each module with the step-size attribute. If imported modules have different precisions, then the highest precision will be used. It might happen that the actual test system cannot provide the required precision, in this case the user is informed and test cases are terminated with an error verdict.

The current value of the clock is retrieved as a float value by the TTCN-3 expression now. Vice versa, the execution of a test component can be suspended until a specific point of time. The operation wait has a float parameter, the execution of the test component will be suspended until the clock reaches this value. Calling the operation wait with a time as parameter that has already passed is considered an error.

The wait operation can be used to send a message at a specific time, see the example in Table 12.5.

Actually there is still some internal processing between the wait operation and the send operation. Some further small delay might occur until the message actually leaves the test system. These delays when sending a message are usually very small. When

Table 12.5   Send message at specific time

```
wait( 37.05 );
pt.send( a_msg );
```

Table 12.6 Retrieving the reception time

```
var float sndTime := 37.05;
var float rcvTime;
const float maxDelay := 0.01;
wait( sndTime );
pt.send( a_msg );
pt.receive( a_reply ) -> timestamp rcvTime;
if ( sndTime + maxDelay < rcvTime) { setverdict( fail ) };
```

receiving a message the delay between receiving a message from the system under test and actually processing the received message in a `receive` statement might be much larger.

When receiving a message from the system under test and enqueuing it, a timestamp is assigned. This timestamp can be retrieved by redirecting it to a variable in a `receive` statement. In the example, shown in Table 12.6, a message is sent at a specified time and the timestamp of enqueuing the reply is retrieved and compared against the sending time. If the reply was enqueued too late the verdict is set to fail.

Redirecting the timestamp to a float variable is possible for all receiving operations of both message and procedure-based communication. Ports where such timestamps are used have to be marked as `realtime` in their definition.

This extension package does not provide a notion of hard real-time, ensuring that specific actions take place at the specified time. But it allows the reception of messages to be measured more accurately. Note that the clock is global for all test components in a test system, whereas timers are local to test components. Without the global clock it would not be possible to suspend execution of two test components until the same moment in time.

12.3   Type Parameterisation

TTCN-3 types can be defined in an open way by using type parameterisation [18]. Type parameterisation is defined in two flavours: Firstly by value parameterisation of types and secondly by allowing the use of types as parameters. The usage of types as parameters can resolve the shortcomings of the `anytype` type discussed previously in Chapter 9.

12.3.1   Value Parameterisation of Types

Values are often used in type definitions to restrict the length of list-like parts or to express range restrictions. If the same type definition shall be used with different restrictions then the same type definition can be repeated again and again with different restrictions. But it is also possible to define a type with value parameters and instantiate the parameters differently. Thereby code replication can be avoided.

The example in Table 12.7 has been modelled on the Mobile Application Part (MAP) protocol [41], it shows how the length of a `record of` type can be defined by a parameter. The type `ProtocolParameters` is a record with two fields, defining an upper

Table 12.7 Value parameterisation of types

```
type record ProtocolParameters {
  integer upperBound,
  integer lowerBound
}
const ProtocolParameters c_XProtocol {
  upperBound := 3,
  lowerBound := 0
}
type charstring Name(ProtocolParameters params)
  length (params.lowerBound .. params.upperBound)
type record Header(ProtocolParameters params) {
  Whatever whatever,
  Name(ProtocolParameters) name
}
```

Table 12.8 Instantiation of value parameterisation of a type

```
type record XPDU {
  Header(XProtocol) header,
  octetstring data
}
```

and a lower bound. A constant of this type provides specific values, which will be used later in the definition of a specific protocol.

The type Name is a character string with length restrictions. In its definition, the length restrictions are left open, the definition refers to fields of the formal parameter. The type Header is another parameterised type, here the formal parameter is just passed on to the type of a field. Such parameter type definitions can be instantiated in other type definitions.

Building on this example in Table 12.8, the definition of the type XPDU is not parameterised itself, but it refers to a parameterised type definition for its fields. Specific values are provided by the actual parameter. In this example the constant XProtocol is used as the actual parameter. Another protocol could reuse the definition of the type Header with different length restrictions.

Collecting the specific values in a constant allows them to be defined in a common place and used in different type definitions consistently.

Even in cases when value parameterisation is not used directly in TTCN-3, it might occur in types imported from other languages. As an example, the MAP protocol has been defined originally in ASN.1 using value parameterisation in ASN.1. Without value parameterisation in TTNC-3 it would not be possible to import these ASN.1 definitions completely into a TTCN-3 test suite.

It should be noted that to avoid confusion with the parameters defined in a signature definition, using value parameterisation for a signature is not allowed.

Table 12.9   Generic protocol definition with type parameter

```
module transport
{
  type record ProtocolDataUnit <in type Payload>
  {
    ProtocolHeader header,
    Payload        payload
  }
}
```

12.3.2   Types as Parameters

Going one step further, it is even possible to use types as parameters of types and other definitions. Types as parameters allow reusable libraries to be defined, where specific types are left open. The shortcomings of using the type anytype, see Section 9.2.1, as a placeholder for arbitrary other types can be avoided.

The example in Table 9.8 can be rewritten with a parameterised type. The type ProtocolDataUnit has a type parameter Payload. A type parameter cannot be changed in a definition, therefore it has to be an **in** parameter. Type parameters are written in angle brackets to distinguish them from value parameters.

In Table 12.9 an example of type parameterisation is shown. In this definition, the type parameter Payload is used to define the type of the field payload.

This type definition can be used in other types but also in behaviour definitions. In Table 12.10 the parameterised type definition is imported. Using the parameterised type definition both a port and a component type are defined, which again have type parameterisation. In these definitions the type parameter is used as an actual parameter of a type in the body of the definition. The port type EchoPort is defined as a port for message-based communication where messages of type ProtocolDataUnit can both be sent and received. Both of these types have a type parameter. Similarly the component type EchoCT is defined as a parameterised type.

The template a_pdu is used to exchange messages of the parameterised data type. This template has again a type parameter Payload. This type parameter is used as actual parameter of the template type – ProtocolDataUnit<template> – as well as the type of the value parameter p_payload of the template. This ensures that the template definition is type correct.

Going one step further a test case with a type parameter is defined. The type parameter Payload is used as a type of the test case parameter p_payload, as parameter of the component type on which the test case is executed, and as parameter of the templates in the test case.

The meaning of a definition with type parameters is provided only when it is instantiated. The usual TTCN-3 semantics is applied to the instantiated definition. This means that the meaning of the test case tc_echo is only provided for instantiations of this test case.

There are two important restrictions on the use of type parameters. Firstly, only definitions inside a TTCN-3 module can have type parameters. But a module itself cannot

Table 12.10 Use of type parameters

```
module echo
{
  import from transport all;

  type port EchoPort <in type Payload> message {
    inout ProtocolDataUnit <Payload> }
  type component EchoCT <in type Payload>{
    EchoPort <Payload> pt }

  template ProtocolDataUnit<Payload> a_pdu
    <in type Payload> ( in Payload p_payload ) :=
    header := c_header,
    payload := p_payload
  }

  testcase tc_echo <in type Payload> (in Payload p_payload)
  runs on EchoCT <Payload>  {
    timer t_guard;
    t_guard.start(1.0);

    pt.send(a_pdu<Payload>(p_payload));
    alt{
      [] pt.receive(a_pdu<Payload>(p_payload)) {setverdict(pass) }
      [] t_guard.timeout { setverdict(fail) }
    }
  }
}
```

have type parameters. There are no different instantiations of a module. Secondly, test cases with type parameters can be called only from within a TTCN-3 test suite. Such test cases cannot be called via the TCI. When test cases with type parameters are called from within a test suite, then the specific instantiation is known and can be analysed before executing the test case. Calling such a test case via the TCI would easily allow a new instantiation to be provided at run time. But such an instantiation might not have been analysed beforehand and actually might be incorrect TTCN-3. This restriction was therefore necessary to enable complete analysis of TTCN-3 code before execution.

Types as parameters are mostly useful to define libraries that can be used in several test suites. In the specific test suites the type parameters of the library would be instantiated, but no new type parameters would be introduced. Therefore the definitions of the actual test cases would not be cluttered with type parameters.

12.4 Behaviour Types

Once defined, functions, altsteps and test cases can be called at different places in a test suite. But the name of the function, altstep or test case has to be written in the TTCN-3 code, and it is not possible to change this without recompiling or reinterpreting the test suite. In many test suites or in large parts thereof there is no need to change this. But there are cases where this is really a limitation.

We showed in previous sections that types as parameters can be used to define libraries of reusable code where certain types have been left open. Similarly, one might want to leave certain behaviour specifications open until the library is really used. A typical example are call-back functions, that are registered to the library and which are called from within the library code. But such call-back functions are not defined within the library code, nor should they be imported to the library. Typically, such a call-back function would be registered by passing it as a parameter to some registration function of the library. This implies that the registration function would have a parameter of a type to denote functions, altsteps or even test cases.

A specific example of such a library could be one for load or stress testing. The library could provide different functions to execute behaviour according to configurable patterns, such as statistical distributions over time, and to collect the results of the executions. The library itself would provide, for example definitions to start individual test behaviours according to statistical distributions. Such a library could be used in load testing different network elements such as a telephony switch or a DNS server. The behaviours executed against the system under test would be quite different in these cases and could be passed as parameters to the library for load testing.

Behaviour types have been introduced as an extension package [17] to TTCN-3. A behaviour type is defined similar to an actual function, altstep or test case, but without behaviour statements. Therefore a behaviour type can be seen as a prototype of a function. The behaviour type is denoting those functions, altsteps and test cases respectively of the test suite that have compatible parameter lists.

In the example in Table 12.11 some types and functions for ordered trees of character strings are defined. The actual ordering routine is left open and passed as a parameter. The function type Smaller is defined as those functions that have two character strings as in parameters and a boolean return value. The function insertOrdered has a parameter cmp of type Smaller. This parameter will be used as the ordering routine when inserting a leaf into the tree.

Such a function value is called on certain parameters by passing them to the new predefined function apply, similar to a start statement. This can be seen in the body of the function insertOrdered in the continuation of the example in Table 12.12.

As a concrete example a test suite could define a number of comparison operators on strings such as shown in Table 12.13.

Table 12.11 Behaviour types for sorted trees of character strings

```
type function Smaller(in charstring a, in charstring b) return boolean;
type Tree {
  charstring e optional,
  Tree       l optional,
  Tree       r optional
}
function insertOrdered(inout Tree tree, in Tree leaf, in Smaller cmp)
{
  //to be completed
}
```

Table 12.12 Calling a parameter of a behaviour type

```
function isEmpty(in Tree t) return boolean {
  return not ispresent(t.e)
}
function insertOrdered(inout Tree tree, in Tree leaf, in Smaller cmp)
{
  if (isempty(tree)) {
      tree := leaf
      } else if ( apply(cmp(tree.e, leaf.e)) ){
      // element in leaf is smaller than element in root of tree
      insertOrdered(tree.left, leaf, cmp)
      } else if ( apply(cmp(leaf.e, tree.e)) ){
      // element in leaf is larger than element in root of tree
      insertOrdered(tree.right, leaf, cmp)
      } else { // do nothing, leaf is already in tree
      }
  }
}
```

Table 12.13 Behaviour values

```
function lexOrder(in charstring a, in charstring b) return boolean {
  // comparison according to lexicographical order
}
function numOrder(in charstring a, in charstring b) return boolean{
  var int a_num := char2int(a);
  var int b_num := char2int(b);
  return ( a < b )
}
function equal(in charstring a, in charstring b) return Boolean {
  return (a == b}
)
...
insertOrdered( t, l, lexOrder);
...
```

Any of the three functions in Table 12.13 can be used as a parameter of the function `insertOrdered`. Note that there is no semantic check whether the functions have a specific behaviour. Certainly, when using the function `equal` as parameter to `insert-Ordered`, the result will not be an ordered tree. Whether a function is an element of a function type is decided by purely syntactical means, that is whether parameters and return types of the function and the function type are compatible. Each function type also has a value **null**, which can be used to denote an undefined behaviour. Also predefined functions of TTCN-3 as well as user-defined external functions are possible values.

The rules for compatibility are rather strict: A function can only be compatible with a function type, and a similar rule holds for altsteps and test cases. Hence no altstep can be compatible with a function type. The parameter list of a function is compatible with the parameter list of a function type if the number and order of parameters are the same. Parameters have to have the same direction, name, kind, type and default. In the case of template parameters potential template restrictions also have to be identical. A function with a return value can be compatible with a function type with the same return type only.

If the function has a `runs on` clause referring to a component type, then the type in the `runs on` clause has to be an extension of the type in the `runs on` clause of the function definition.

In the example before we used a parameter of a function type. But it is also possible to define variables, constants and templates of function types. Also the return type of a function can again be a function type.

When assigning a function to a parameter or variable then as usual in TTNC-3 the compatibility is checked according to the rules explained before. In addition to assigning a function to parameters or variables, or applying it to arguments, the only operations allowed on functions are checks for equality and non-equality.

When assigning a function to a parameter or variable, then the compatibility of the component types in the `runs on` clause are also checked. The function assigned must be compatible with the type of the parameter or variable, which is the case when the component type of the function is an extension of the component type of the function type. Actually applying such a function then might cause problems because the function could be executed on the component type of the function type. For this to work, this component type should be an extension of the type of the function, but the compatibility rule is the other way round. This problem typically occurs in library code using function types. This is solved by the specific `runs on` clause `runs on self` that can be used in function type definitions. This clause means that the function type itself does not have its own component type. Instead, wherever in the TTCN-3 test suite a compatibility check against this function type is performed, the component type of the enclosing behaviour is used instead. For further detail on this topic the reader is referred to [17].

Most of this section has been written with a focus on functions. But similarly it is possible to define and use types of altsteps and test cases. Note that for test cases the usage of `runs on self` is not allowed.

Type parameterisation and behaviour types have been described here independently of each other. Actually it is possible to define types of functions with type parameters. For example, instead of defining functions for ordered trees of character strings, it is also possible to define functions and types for ordered trees of arbitrary types. Even in this combined case it is still possible to perform the usual syntactic and semantic checks of TTCN-3 by static analysis before actually executing the tests.

12.5 Summary

In this chapter we have explained extension packages as a means to extend TTCN-3 with new concepts. Four different extension packages have been defined so far and are explained in this book.

Extension packages have been introduced only lately to TTCN-3 and limited experience could be gained so far. Time will show whether this approach to extend TTCN-3 is a feasible one and whether the concepts introduced by the extension packages gain acceptance in the field.

13

TTCN-3 Test Systems in Practice

The development of a complete test system involves more than just writing the Testing and Test Control Notation Version 3 (TTCN-3) code that describes the behaviour. In this chapter, we look in more detail at the overall test system, which has already been presented briefly in our very first TTCN-3 example. We will first review the overall structure of a TTCN-3 test system and then present its most important entities in more detail. We will introduce the roles of these entities, how they work internally, and how they interact. These concepts will be explained using an elementary TTCN-3 test case similar to the test cases that we have already studied in Chapter 2. In addition, we will discuss the interfaces via which the interactions between test system entities take place. These are the standardised TTCN-3 Runtime Interface (TRI) and TTCN-3 Control Interface (TCI).

Although this chapter mainly discusses general test system implementation aspects, we will again use the Domain Name System (DNS) example, which was first introduced in Chapter 2 for some more explicit explanations. We restrict ourselves in this chapter to explain the implementation issues for test systems according to the main parts of the TTCN-3 language parts. This means we do not consider the extension packages for test configurations and for real-time testing here. The extension packages contain the definition of operations to implement functionality similar to the one presented here, but they are not described here for the sake of simplicity.

13.1 The Anatomy of a TTCN-3 Test System

A TTCN-3 test system can be conceptually defined as a collection of different test system entities, which interact with each other during a test suite execution. Figure 13.1 gives a schematic view of this architecture, which consists of three dominant layers. A central layer, the TTCN-3 Executable (TE), handles the execution of TTCN-3 statements. For its operation, the TE depends on a number of services that are provided by the other two main layers. The Test Management and Control (TMC) entity is responsible for aspects like interfacing to the test system user, the encoding and decoding of data, logging and aspects that deal with distributed execution. These services are provided by the

An Introduction to TTCN-3, Second Edition.
Colin Willcock, Thomas Deiß, Stephan Tobies, Stefan Keil, Federico Engler and Stephan Schulz.
© 2011 John Wiley & Sons, Ltd. Published 2011 by John Wiley & Sons, Ltd.

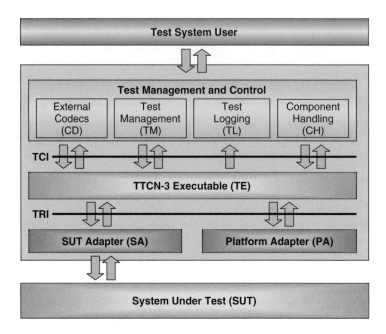

Figure 13.1   Conceptual model of a TTCN-3 test system.

Test Management (TM), External Codecs (CD), Test Logging (TL) and Component Handling (CH) entities, respectively. For the interfacing towards the system under test (SUT) and towards the actual test system operating system, the TE uses services provided by the two adapters SUT Adapter (SA) and Platform Adapter (PA). Communication with the central entity takes place via the standardised TCI [13] and TRI [12, 42]. Both these interfaces are defined as a set of operations, which are called by one entity and implemented within another.

This standardised separation of tasks into different entities makes TTCN-3 test system implementations very flexible. It enables, for example the reuse of SUT interface code in different test suites. It also enables the execution of the same test suite in different testing phases. In practice, TTCN-3 tools provide users with (default) implementations for some of these entities. Among this class are the TM and CH entity, which we will not discuss in this book. Other entities, namely the CD, SA and PA entities, are usually not provided as they cover aspects of a test system, which are either test suite or SUT specific. These entities typically need to be implemented by users and will be discussed in the following sections of this chapter.

Note that in this chapter we have simplified our discussion of entity interactions via the TRI and TCI by using only operation names. More details about these operations, for example the exact definition of their parameters, can be found in the standard documents [12] and [13]. Both standard documents define the operations abstractly in IDL [27], additionally they provide mappings from IDL to specific programming languages, namely C, C++ and Java. If a user has to implement some of the operations at all, then the user does not have to cope with the abstract interfaces, but may already use the interfaces in a specific programming language. An additional mapping to XML is defined for the logging operations defined at the TCI-TL interface. The corresponding XML schema can serve as a common format for logs. Large organizations using different TTCN-3 tools may still use the same tools for post-processing TTCN-3 logs.

13.1.1   The TTCN-3 Executable

The TE is at the heart of a TTCN-3 test system. Note that the name 'TTCN-3 Executable' does not necessarily imply that this entity comprises a separate executable program. Indeed, most TTCN-3 tools treat entities as libraries that are then combined with the TE implementation to obtain an executable test system. The name 'TTCN-3 Executable' shall only indicate that this entity is responsible for the execution of the TTCN-3 code.

Both interface standards, TRI and TCI, intentionally make only a few assumptions about the actual implementation of the TE. It must consist of a suitable representation of your TTCN-3 test suite plus some mechanisms to execute this code as specified by the TTCN-3 core language standard. These mechanisms, which are usually referred to as the Run Time System (RTS), must then use the services provided by the other test system entities in its execution of the TTCN-3 test suite. The interface standards, however, make no assumption about the programming languages used to implement this as well as the other entities. This gives TTCN-3 tool vendors a great amount of flexibility in their tool implementations.

The good news is that a tool's RTS implements all the advanced aspects of TTCN-3 semantics for you, for example concurrent test components, snapshots, verdict handling, memory management, dynamic type checking and so on. Therefore, the TE provided by a tool will generally allow us to concentrate on the essentials of testing.

13.2   Test System Execution of a Simple Test Case

Before we give a more detailed explanation of the entities that support the TE in the execution of the TTCN-3 code, we will sketch the execution of a simple test case, as seen via the TRI and TCI interfaces. For this, we will return to the DNS test case example, which we presented in the introductory chapter. The code is repeated and slightly modified in Table 13.1. Type and template definitions will be provided in the following sections when we discuss encoding and decoding of values in more detail.

To start, let us take a closer look at the various things that will happen within the TTCN-3 test system when this code executes. Figure 13.2 shows a simplified view of the operation invocations that will take place.

13.2.1   Test System and Test Case Initialisation

During the initialisation phase of the test system, which is shown in Figure 13.3, the TE invokes the TRI operations `triResetSA` and `triResetPA`, which are provided by the SA and PA entity, respectively. Once these entities have indicated their successful initialisation, the TE starts executing the test suite control part. When the TTCN-3 `execute` statement is encountered, the TE invokes the `triExecuteTestCase` operation on the

Table 13.1   The example DNS server test case

```
testcase tc_resolve() runs on DNSClient system DNSClient {
  timer t_replyTimer;

  map( self:pt_clientPort, system:pt_serverPort );
  pt_clientPort.send( a_NokiaQuestion );
  t_replyTimer.start( 20.0 );
  alt {
    // Handle the case when the expected answer comes in.
    [] pt_clientPort.receive( a_NokiaAnswer ) {
        setverdict( pass );
      }
    // Handle the case when an unexpected answers come in.
    [] pt_clientPort.receive {
        setverdict( fail );
      }
    // Handle the case when no answer comes in.
    [] t_replyTimer.timeout {
        setverdict( inconc );
      }
  }
  t_replyTimer.stop;
  unmap( self:pt_clientPort, system:serverPort );
}

control {
  execute( tc_resolve() );
}
```

SA to inform it that a new test case is about to be started. This allows the SA to prepare its communication facilities to be used for communication with the SUT. In our DNS example, this could, for example mean the initialisation of the User Datagram Protocol (UDP) and IP protocol layer in the operating system.

13.2.2   Preparation of Communication Channels towards the SUT

The SA also implements the `triMap` operation (and its counterpart, the `triUnmap`), which is called by the TE upon executing a `map` statement in a TTCN-3 test suite; this is shown in Figure 13.4. According to the TRI standard, this operation should be used to prepare a SUT communication interface for the interaction with the SUT. From the TE's point of view, successful completion of the `triMap` operation enables a test component to communicate with the SUT. TTCN-3 has been designed to allow testing of arbitrary systems and makes no assumptions about how the communication with the SUT will actually be established. This flexibility has a price, namely, that during test system development it is necessary to define a mapping from the relatively abstract TRI operations onto the concrete operations that are needed to communicate with the SUT.

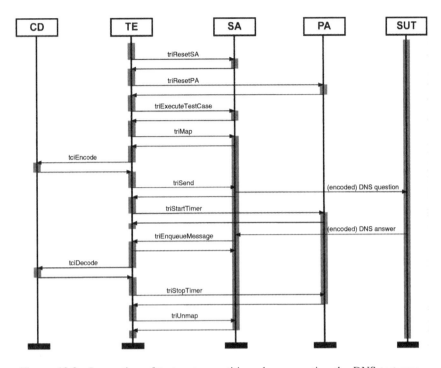

Figure 13.2   Interaction of test system entities when executing the DNS test case.

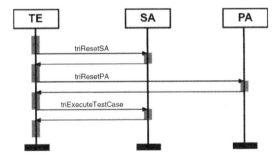

Figure 13.3   Interactions during DNS test system and test case initialisation.

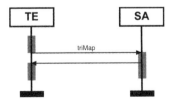

Figure 13.4   Interaction performed to set up UDP IP transport.

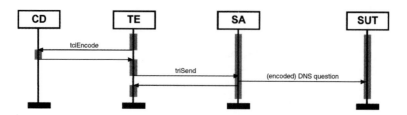

Figure 13.5   Interactions performed to send a message to the SUT.

For our DNS test case, for example invocation of `triMap` could trigger the allocation of a UDP socket and port through which DNS server messages should be received.

13.2.3   Handling of Communication towards the SUT

The next step in our test case execution, which is shown in Figure 13.5, is the sending of the message `a_NokiaQuestion` to the SUT. To achieve this, the message will first have to be encoded from a structured TTCN-3 value into a form, that is accepted by the SUT. Second, this message has to be dispatched to the SUT. In the same manner that TTCN-3 does not make assumptions about the mechanism, that is used to communicate with the SUT, there is also no assumption about the way that messages have to be encoded for the SUT to understand them. This means that the concrete encoding will have to be implemented during test system development.

Encoding and decoding services are provided by the CD entity, which is attached to the TE via the TCI. It implements the operation `tciEncode` and its counterpart operation `tciDecode`. It encodes a requested TTCN-3 message value and then passes it back to the TE as a binary string. This binary string is then passed on to the SA via the `triSend` operation. The latter invocation also contains information about the sending test component and the information on which test system interface (TSI) port the message has been sent. It is now the responsibility of the SA to transmit the message to the SUT. In our DNS example, `tciEncode` would turn the structured TTCN-3 `Dns-Message` value into its encoded counterpart following the encoding rules specified in the DNS protocol standard [29]. Even if a textual encoding is used such as, for example Session Initiation Protocol (SIP) messages, the TE will use internally binary strings for encoded values.

13.2.4   Starting of TTCN-3 Timers

Once the message has been sent to the SUT, a timer is started in our test case to guard execution of the subsequent TTCN-3 `alt` statement; this is shown in Figure 13.6. The PA is responsible for providing timer services to the SA. The starting of the timer is requested via the TRI interface by a call of the operation `triStartTimer`. This call specifies the duration of the timer and a handle that shall be used to identify the timer in future communication between TE and PA. Note that this handle is not the TTCN-3 timer name, that is in our DNS test case `t_replyTimer`, but instead an opaque identifier

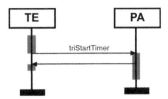

Figure 13.6   Interaction for starting a timer.

selected by the TE. This allows a generic implementation of timers in the PA, as we will see in our later section on the PA.

When the timer has been successfully started, execution of our DNS test case inside the TE will proceed to the `alt` statement, which contains different alternatives to deal with the different possible reactions from the SUT. At this point, the execution of the test case first checks if any DNS message has arrived or the timer has expired. As neither has happened at this point, the TE will block the further execution of the DNS test case.

13.2.5   Handling Incoming Communication from the SUT

If the DNS server accepts our DNS message, it sends a DNS answer message back to the UDP port from where the corresponding query originated. This message is received by the SA, which will forward it via the TRI to the responsible test component inside the TE by invoking the `triEnqueueMsg` operation; this is shown in Figure 13.7. Like its counterpart `triSend`, the `triEnqueueMsg` operation always passes messages in an encoded form.

Inside the TE, the arrival of the message will trigger a new evaluation of the `alt` statement where the first alternative calls for a matching attempt of the received, encoded message against the specific DNS message `a_NokiaAnswer`. For this, the encoded message first has to be decoded into a structured TTCN-3 value. This service is provided via the TCI by the CD entity, which implements the `tciDecode` operation. In addition to the encoded message, the TE also must specify the assumed type of the message, that is the decoding hypothesis. This decoding hypothesis will be used within the CD to select a decoding mechanism. This decoding hypothesis is needed because the CD may provide services for more than one protocol.

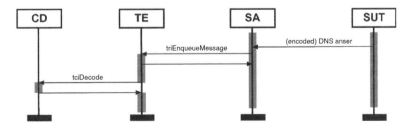

Figure 13.7   Interactions after reception of a message from the SUT.

In the case of our DNS example, the decoder would check if our received message is a correctly encoded DNS message. A successful check will then create a TTCN-3 DnsMessage value of the message and return it to the TE. This value will be used in a template match attempt inside the TE, which will then cause selection or rejection of the currently considered alternative. Failure to decode the message is also reported to the TE and will cause rejection of the currently considered alternative.

13.2.6 Handling Timeouts and Stopping of Timers

Assuming that our received DNS message matches, the execution of our test case will then proceed to the timer stop statement. This will cause the TE to request stopping of the timer that was previously started in the PA. This is accomplished by calling the triStopTimer operation via the TRI as shown in Figure 13.8. The operation should succeed even for previously stopped or already timed-out timers and will allow the PA to discard the timer.

If no answer has arrived within the specified 20-second duration, the timer will expire in the PA. The PA will indicate this via the TRI by calling the triTimeout operation as shown in Figure 13.9, specifying the handle of the timer that has expired. Like a new incoming message, a newly occurred timeout will also cause a new evaluation of the alt statement. The alt statement would then select the third alternative, causing the inconclusive verdict to be set. Note that even in this case the TE would attempt to stop the timer as discussed before.

13.2.7 Teardown of Communication Channels towards the SUT

Before the test case stops, it will unmap the Main Test Component's (MTCs) port pt_clientPort from the TSI, which will cause the operation triUnmap to be

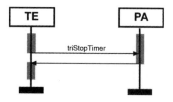

Figure 13.8 Interaction to stop a timer.

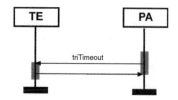

Figure 13.9 Interaction in case of a timeout.

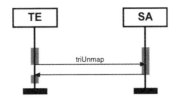

Figure 13.10   Interaction to tear down a communication channel.

called via the TRI, which is shown in Figure 13.10. This allows the SA to tear down the communication channels that have been established by the corresponding `triMap` invocation.

In our DNS example, this would mean releasing the UDP socket and freeing the UDP port that was allocated for communication with the DNS server. Once a port has been unmapped, no more messages can be sent to the SUT via this port, and no messages from the SUT can be reported to the TE by calls to `triEnqueueMsg`. Finally, the operation `triEndTestcase` is called by the TE. It allows the SA to free resources or to close communication ports with the SUT.

This concludes the example test case execution.

13.3   More about the SUT Adapter

We have seen that the SA's role is to provide the means for communication between the TE and the SUT and to bridge the gap between the (abstract) TRI communication primitives and real communication mechanisms employed by the SUT. Because of the abstract nature of TTCN-3's communication mechanism, it will usually be up to the test system developers to implement this particular test system entity.

The main task of the SA is to add transport information to encoded messages or calls sent by the TE and send them to the SUT. Conversely, it must be able to receive messages or procedure calls sent by the SUT during test case execution, extract from them the data relevant to the test suite (that is, strip off transport information) and then forward this encoded data to the TE. All TRI operation implementations in the SA have to be re-entrant because concurrently executing test components may simultaneously invoke these operations.

13.3.1   Execution Threads in the SA

Since an SA must be able to receive messages from the SUT at any time during test case execution, a test system usually requires a concurrent design for the SA. With concurrent design, one or more *separate* execution threads can deal with incoming messages, incoming remote procedure calls (RPCs) or replies to previously invoked remote procedures. These will then be passed to the TE using the TRI operations `triEnqueueMsg` or its counterparts for procedure-based communication.

Note that our previous example in Figure 13.2 shows that the SA is always active after the invocation of the `triMap` operation. This has been intentionally shown this way to symbolise that a separate thread within the SA is always checking for the arrival of any messages from the DNS Server.

13.3.2 Management of TRI Information

Each TRI communication operation, regardless of whether it is from or towards the TE, must specify which test component port is involved in this communication. In SA implementations with multiple test components, this information can be used to determine which connection, for example UDP socket, is to be used for sending the data. Conversely, each incoming message will have to be addressed to a test component in the TE. This will usually require maintaining some form of state in the SA that binds component ports to SA communication ports that connect to the SUT. Typically, this state will be created (and respectively destroyed) by the operations `triMap` and `triUnmap` and is used in the implementation of the TRI communication operations.

13.3.3 Procedure-Based Communication with the SUT

In our previous test case execution example, we only covered message-based communication, but the TRI also contains operations for procedure-based communication. Such transport service must also be implemented in the SA using the `triEnqueueCall`, `triReply` and `triRaise` operations. These operations then need to deliver the data to the SUT. Depending on the underlying RPC mechanism, decoding of the encoded parameter values may be required prior to making the call. For the handling of incoming calls from the SUT, the TRI offers the enqueuing operations for all procedure events, for example `triEnqueueCall`, `triEnqueueReply` and `triEnqueueException`. These operations are implemented within the TE.

It is important to keep in mind that the implementation of TRI communication operations in the SA *must not* implement the TTCN-3 semantics. A `triEnqueueCall` operation, for example must not block until the SUT has replied. Instead, it must return immediately after dispatching the call. Depending on the underlying RPC mechanism, this will require the delegation of the communication to a concurrent thread or using an asynchronous call mechanism. This has the advantage that the correct implementation of TTCN-3 semantics is isolated in the TE and therefore not of concern to the user.

13.3.4 Dynamic SUT Adapter Configuration

An important aspect that we have intentionally skipped in our initial test execution example is the configuration of the transport mechanism, for example the address of the SUT or quality of service parameters to be used in the communication with the SUT. A number of different solutions are possible in order to handle these aspects in an SA implementation. The easiest solution is to simply hard-code such information into the SA. In the context of our DNS example, this would mean that all DNS messages are simply sent to a fixed SUT IP address and UDP port.

Obviously, such a design is very inflexible and would tie our test system to one specific SUT or would at least require some form of re-compilation when the SUT moves to a new address. Alternatively, external configuration files that are read by the SA can be used to achieve a limited form of flexibility of SA configuration. From our experience, the most flexible approach to configure the SA is by exchanging configuration messages with it. For this purpose, a distinguished TSI port is set aside and not used for sending messages to the SUT. Rather, the port is used to send configuration messages directly to the SA implementation. These messages carry the necessary configuration information in a suitable configuration protocol and are decoded by the SA to control its operation. Using this approach, it becomes possible to even re-configure the SA during or between test cases and to treat configuration errors flexibly.

13.3.5 Distributed SUT Adapter Implementations

Finally, we want to point out that the monolithic representation of an SA, as shown previously in Figure 13.1, does not necessarily imply that an SA must be implemented within the test system executable or is necessarily tied to the same computer. In large telecom test systems, for example the implementation of the protocol stack, necessary to communicate with the SUT may indeed be the most complex part of the whole test system, and the emulation of the involved protocol layers will require substantial processing power.

In cases like this, it is advisable to split the SA implementation from the TE into a separate executable, potentially even running it on a separate machine. The same advice holds for cases where a large amount of different communication interfaces have to be supported by a test system implementation. In this case, it might even be useful to have several separate executables that constitute the SA. Distributed SAs can be integrated with a test system by simply implementing the TRI interface of the SA as a thin proxy layer that dispatches the TRI operations to the different SAs. Each of these SAs could then, for example implement one communication interface, as shown in Figure 13.11.

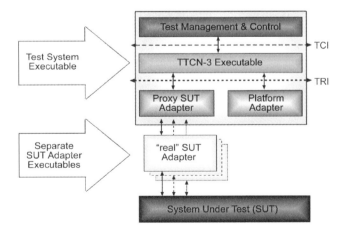

Figure 13.11 Test system with distributed SA implementation.

13.4 More about the Platform Adapter

The PA implements those test system adaptation aspects that are not directly related to the interaction with the SUT. It implements the model of time to be used during the execution of TTCN-3 as well as external functions. The reason these aspects have been isolated from the TE entity is to make a test system implementation more independent from the underlying operating system and time model, as well as the concrete implementation language used to implement external functions.

13.4.1 TRI Timing Operations

It is the PA's responsibility to provide a coherent notion of time for the execution of the test system. For this purpose, the PA does not have to distinguish between the different forms of timers that occur in the TTCN-3 code, like test case timers, timers associated with blocking `call` statements or explicit timers. All that matters to the PA is the (opaque) timer identifier, which is created by the TE to name each particular timer instance.

In addition to the previously introduced operations, the PA must also implement the access operations `triReadTimer` and `triIsTimerRunning`. These operations are used to query a timer's state in the PA. Note that, like in the case of the SA implementation, the PA will also usually require a concurrent implementation because of the asynchronous nature of timeouts. It will also be necessary to implement all PA operations in a re-entrant manner because multiple components may access timers concurrently.

13.4.2 Non-Real-Time Implementation

The TRI standard does not make any assumptions about the underlying notion of time, that is implemented by the PA. Hence, a PA is not required to reflect real, wall-clock time. Instead, the timer implementation can also be done in such a manner that it integrates a test system with a SUT debugging or simulation environment. When a SUT is run in debugging mode, a wall-clock time implementation would cause test case execution failures due to timeouts, while the SUT is halted by the user at break points or when simulation of the SUT is computationally expensive and requires more time for certain steps than allowed for in the test suite. This can be prevented by letting the debugger or simulation environment control the passing of time, for example by halting time when the execution is halted in the debugger, or during the execution of simulation steps.

13.4.3 External Functions

Some parts of a test system implementation will rely on functionality, that is only available outside the test system and not concerned with communication towards the SUT. Examples of this are operating system functionality like file system access, database integration, library implementations for mathematical functions, or interfacing with test equipment like data generators or protocol analyzers. Access to such external functionality can be provided by means of external functions. Inside the TTCN-3 code, external function calls are indistinguishable from the invocation of ordinary TTCN-3 functions. However, their invocation will not be handled inside the TE, but instead the `triExecuteExternalFunction` operation will be invoked within the PA. This invocation names the function to be executed and specifies the (encoded) parameter list. It will then be the PA's responsibility to take the appropriate action to execute the concrete implementation of the external function.

External functions are a powerful construct but have to be used with care. Incorrect external function implementations may cause hard-to-track test execution errors for TTCN-3 users and writers, and are hard to identify from within the TTCN-3 code. Also, external function implementations require more work than test system developers may anticipate at a first glance: `in` and `inout` parameters for external functions are passed in encoded form and will have to be decoded before being used in the external function. Conversely, return values, `out` and `inout` parameters have to be encoded by the external function upon returning from its invocation. This requires codec implementations for the parameter types both in the test system and the external function implementation and will also consume non-negligible processing time. Such a codec is necessary to perform conversion between the value representation used by the TE and that used by the external function implementation.

13.5   More about External Codecs

Many protocols or software systems use their own proprietary way of encoding information. The user will have to provide an implementation for any encoding scheme, that is not supported by the TTCN-3 tool. For example, both the DNS and SIP protocols we have shown in this book use proprietary encoding schemes, and therefore will require user-defined codecs. Only when the SUT communicates exclusively using standardised encoding schemes like Abstract Syntax Notation One (ASN.1) Basic Encoding Rules (BERs) or Packed Encoding Rules (PERs) is it possible that a TTCN-3 tool may already offer an implementation of the required codecs. In general, it is the CD entity that has the responsibility to perform both the encoding between the value representation used in the TE and the format expected by the SUT and the decoding in the opposite direction.

13.5.1   Access to the TTCN-3 Values

For both encoding and decoding of values, the CD must deal with the TTCN-3 values. During encoding, a value has to be inspected and dissected to generate the binary representation mandated by the encoding scheme. During decoding, a new value has to be created on the basis of its encoding received from the SUT. This requires the CD to be able to manipulate values inside the TE. Clearly, the representation of values inside the TE will vary between TTCN-3 tool implementations. The TCI standard defines abstract `Type` and `Value` interfaces, which decouple the CD from the concrete tool's value representation. On the basis of these interfaces, codec implementations (or even codec generation tools) can be implemented in a tool-independent manner.

13.5.2   Encoder Implementation

An implementation of the encode operation is typically rather simple. After inspecting the type of the TTCN-3 value to be encoded, a suitable encoding mechanism will be invoked, which must construct and return the encoded form as a binary string based on the chosen encoding scheme. Of course, depending on the encoding scheme, more or less effort will be required to actually generate this encoded form. Encoding a value will usually require a systematic traversal of its structure, which can be perceived as a labelled tree, and the appropriate assembly of the encodings of the sub-trees that form the value.

Traversal and inspection of the value tree is done via the TCI Value interface, which provides a number of access operations for values of all basic as well as user-defined TTCN-3 types. For example, the getField operation retrieves a specific field value from a record or set value, the getBoolean returns the actual value assigned to a TTCN-3 boolean value and so on. For the construction of the encoded data, which is returned by the encode operation, the TCI only specifies the concrete value structure, which is the same as that used in the TriMessageType type. Neither the TRI nor the TCI specify operations for the construction of such values. Therefore, the CD must implement such operations itself.

In the context of our DNS message example, we are dealing with the types and values shown in Table 13.2. During encoding, the TE invokes the encode operation in the CD and passes the template a_NokiaQuestion as the DnsMessage value to be encoded. To encode this value, the encoder would first retrieve the identifier field of the Dns-Message value using the getField operation and then extract the value assigned to this field, that is 12345, using the getInt operation. The DNS standard [29] mandates encoding this identifier value as 16 bits in network byte ordering. Therefore, the encoder would begin the encoded message with the bytes 0x3039. It would then continue by

Table 13.2   TTCN-3 type definitions for the DNS message and a value

```
type integer      Identification( 0..65535 );
type enumerated   MessageKind { e_question, e_answer };
type charstring   Question;
type charstring   Answer;

type record DnsMessage {
  Identification   identification,
  MessageKind      messageKind,
  Question         question,
  Answer           answer optional
}

template DNSMessage a_NokiaQuestion := {
  identification    := 12345,
  messageKind       := e_question,
  question          := "www.nokia.com",
  answer            := omit
}
```

accessing the `messageKind` field of this record, extract the `e_question` enumeration value, append a zero bit to the encoded data and then continue this process for the rest of the message.

13.5.3   Decoder Implementation

An implementation of the `decode` operation attempts to construct a TTCN-3 value, based on the expected message type and the encoded data. The first task will thus usually be an inspection of the expected type and the selection of the decoder implementation, that is responsible for decoding this particular type. This decoder would then attempt to build the value structure during a detailed examination of the encoded data. Construction of the value again uses the abstract TCI `Value` interface.

In our DNS example, when called with decoding hypothesis `DnsMessage`, the decoder would start building an empty `DnsMessage` value by calling the `newInstance` operation on the `Type` interface. The decoder would then start inspection of the received message and attempt to decode the first 16 bit as an integer in network byte order. The result of this decoding would then be stored in a newly created instance of `Identifier` using the `setInt` operation. Finally, this value would be used to set the `identifier` field of the `DnsMessage` using the `setField` operation. In the same manner, the remainder of the encoded message would be inspected and all fields of the `DnsMessage` value filled in. The decoding is completed once the encoded message has been completely inspected and all fields of the `DnsMessage` value are set.

Should the decoder encounter some illegal encoding or missing information in the encoded data during the construction of the value, this is a decoding error, which is reported back to the TE by returning from the `decode` operation with an empty decoded value. Otherwise, the result of the successful decoding is passed back.

13.5.4   Advanced Aspects of Codec Implementations

The handling of decoding of subtyped values requires special attention because the set of allowed values is currently not discernible through the `Type` or `Value` interface. Still, the CD will need to check if the value restriction specified by a TTCN-3 type is met *before* the decoded value is set in the TE. An example of such a problematic case would be when an `integer` value is decoded that lies outside the admissible values for a field, for example decoding a negative number to be put into the `identifier` field of a `DnsMessage` value. These cases need to be treated as a decoding error because any attempt to set an incorrect value via the `Value` interface will cause a test case error and hence the termination of the test execution.

This means that there will have to be a very close resemblance between the modelling of message types in the TTCN-3 code and the actual codec implementation. In particular, it must be ensured that a decoder will not successfully decode a message that cannot be represented inside the expected subtype constraints within the TE. In the case of the `identifier` field of the `DnsMessage`, this close resemblance is automatically present because any unsigned 16-bit integer value lies within the admissible range of the `Identifier` type.

Unlike the implementations of SA and PA, it is not necessary for encoders and decoders to execute in parallel to the TE. This is because when the TE invokes the `encode` or `decode` operation, it is prepared to wait for the completion of the encoding or decoding process. Implementations will have to be re-entrant, though, because concurrently executing test components may lead to the simultaneous invocation of `encode` and `decode`.

Similar to our previous discussion of SA and PA implementation, the TCI does not make assumptions about the concrete nature of a CD implementation. For example hard-wiring separate encoders or decoders for specific types is not required, as we have suggested here when discussing codecs for our DNS example. It is also possible to implement more generic codecs, which perform encoding and decoding of values based on a type's structure (as it is possible for BER or PER encoders) or on user-specified meta-information like encoding attributes. We have already sketched this approach in Section 7.5.4.

13.6   Documentation Comments

Comments in TTCN-3 code can be written either as block or line comments. Part 10 of the standard additionally defines a set of specific comment formats for different kinds of TTCN-3 definitions. Such comment formats allow the most relevant part of TTCN-3 definitions to be extracted together with the comments. As an example, the name of a function, its parameters and its return values together with the comments describing the parameters can be extracted. Such extracted information can serve as an interface description of TTCN-3 libraries as they are also common in many programming languages.

As an example, we apply these documentation comments to the function `f_checkHostName` defined in Table 3.8. Documentation comments can be written as block or line comments as any other comment. In the example in Table 13.3 block comments are used. Inside the documentation comment two documentation tags are used, `@param` to describe a parameter and `@return` to describe the return value.

In addition to the `@param` and `@return` tags shown before further documentation tags are defined to document version information, cross references, verdicts of test cases and so on. For the complete list of documentation tags refer to [16]. The text after the documentation tags can be arbitrary freetext, some basic formatting hints to external tools can be given in an HTML like syntax. These formatting hints can be used to emphasise text or to define lists within the comments.

Table 13.3   Documenting a function

```
/*******************************************************************************
 ** @param charstring p_name name to be checked                              **
 ** @return boolean flag, true if name does not contain dots or if there     **
 **                       are more than 63 characters between dots           **
 ******************************************************************************/
function f_checkHostName( in charstring p_name ) return boolean {
  // function body
}
```

13.7 Summary

The composition of a TTCN-3 test system has been defined by ETSI in the TRI and TCI standards. Among the benefits of these well-accepted TTCN-3 test system standards is the possibility of creating SUT as well as TTCN-3 tool independent test system implementations. They also allow the easy integration of TTCN-3 test systems into other tools used, for example for TM.

Each standard first abstractly defines the interface as a collection of operations and then offers mappings of this definition to concrete implementation languages. These interfaces assume a set of test system entities. In this chapter, we have discussed the most important of these entities that are needed to be able to build and run a TTCN-3 test system against a real SUT, namely, the TE, SA, PA and CD. These entities implement communication interfaces to the SUT, external functions, timing as well as encoding and decoding of TTCN-3 values into binary strings.

In a TTCN-3 testing project, the implementation of these entities needs to be considered either prior to or in parallel with the writing of TTCN-3 code. Careful design of these entities will be required to address issues like re-entrant code and asynchronous behaviour that takes place inside these entities. For the remaining test-system entities, TE, TM, CH and TL TTCN-3 tools provide default implementations, which suffice for ordinary test systems.

14

Frameworks[1]

When developing large industrial test suites with multiple test developers, the ability to write consistent, maintainable code becomes very important. Parallel test development also often leads to accidental re-implementation of functionality. In addition, we often encounter situations in testing quite complex interfaces where messages can have a vast amount of parameters that need to be handled in every test. Our SIP examples in the previous chapters have already given a first indication and our LTE example that we will consider in Chapter 16 will show even more the full industrial reality of the possible level of information complexity. Let us keep in mind that SIP and LTE are just two examples and there are many other protocols, especially in the telecom and internet domain.

Test frameworks have been inspired from classic software development practice and try to provide answers to all of these issues. They also allow us to organise more efficiently test suite implementation. Typically such a framework is created by one or more expert test developers. Once the framework is available the bulk of test developers can then focus on testing objectives and functionality to be tested by implementing their code based on these frameworks – not needing to worry about every detail of each and every interface.

14.1 Frameworks and Test Suites

TTCN-3 test implementation based on frameworks can be simply understood as a structured specification of tests with four different layers and based on reusable testing artifacts – often called TTCN-3 libraries. These libraries hide the detail and complexity from test developers, for example default handling for required message parameters or test component synchronisation. The lowest layer with the highest potential of reuse includes the definitions independent of a particular test suite or interface. The second layer collects interface or protocol specific TTCN-3 definitions, for example SIP, DNS or IPV6 specific type, template or behavioural function definitions. These definitions are, if designed well, re-usable across multiple test suites or even types of testing. The

---

[1] TTCN-3 coding and various phrases/diagrams/tables in this chapter are reproduced by kind permission of © European Telecommunications Standards Institute 2008.

An Introduction to TTCN-3, Second Edition.
Colin Willcock, Thomas Deiß, Stephan Tobies, Stefan Keil, Federico Engler and Stephan Schulz.
© 2011 John Wiley & Sons, Ltd. Published 2011 by John Wiley & Sons, Ltd.

third layer collects artefacts within a specific test suite such as test component types, test configuration functions, preambles and postambles. These three layers constitute the framework for test specification. Finally, the top layer contains everything associated with the actual test case definitions which are specific to one particular test suite, such as the control part, test selection functions, test case statements and functions that describe parallel test component behaviour. This top layer finally makes concrete use of the definitions from the framework.

Frameworks have today established themselves in standardised test specifications at ETSI as well as other standardization organisations and industry. Examples include the ETSI IMS automated interoperability testing framework [43], OMA BCAST test suite, the WiMax Network Interoperability test suite [4] or 3GPPs LTE test suite [44] (which will be reviewed Chapter 16). We will base the examples in this chapter on the IPv6 framework [45].

14.2   TTCN-3 Libraries

A TTCN-3 library is a collection of one or more TTCN-3 modules. These modules provide a number of TTCN-3 definitions, which are either independent of other TTCN-3 code, that is self-contained, or based on one or more other TTCN-3 libraries. It is a tool dependent issue if TTCN-3 libraries can be integrated into a test suite in a pre-compiled form or as regular TTCN-3 source code.

At a minimum TTCN-3 libraries must define type and value information. This may include data structures which define the message content of a particular protocol, constants definitions like protocol specific timing values, module parameters or port and component type definitions. Component type definitions are a key element to achieving reusability of behaviour. In order to make them as reusable as possible they should be defined as lean as possible, for example exclude ports for adapter configuration. Test suite component types can then simply be defined as extensions of such library component types. Since the extension naturally ensures component type compatibility of these component types, behavioural functions specified in the library can be invoked from test suite specific code.

Behavioural functions usually define only atomic message exchanges, for example a single send statement followed by the required number of receive statements. In these functions the manipulation of component verdicts is a delicate issue. In the majority of cases, we wish the same behaviour, that is unexpected messages lead to failures, but not always. This means that the setting of negative verdicts can harm the ability to reuse a function. One way to resolve this problem is to use special parameters to indicate the proper verdict setting in such cases. Other useful behavioural definitions are altstep definitions that can then be invoked from test suite specific code to ignore refresh or heartbeat messages.

The main reason for defining a template in a library is its use in one of the behavioural functions. In general, it is advisable to parameterise library templates to allow customisation of templates from the invoking code. Libraries can also be used to define dummy and base templates which set or handle the most important information elements. These templates can then be customised by creating modified template definitions in the test suite specific code [43].

Documentation of library TTCN-3 code is also an important issue, for example using the standardised TTCN-3 documentation tags. Users of libraries should be able to use definitions, for example functions without studying any TTCN-3 code.

14.3 Design of Frameworks

TTCN-3 test suite implementation frameworks are basically a set of rules and guidelines on how the code is structured and implemented. Frameworks are usually developed and customised for each application area. A good framework should include:

- Naming conventions.
- Exhaustive list and illustration of all possible test configurations.
- Methodology for test development based on the framework or libraries including mapping of major application area concepts to TTCN-3 definitions.
- Template and function design guidelines.

14.4 Example: the IPv6 Testing Framework

ETSI test suites are developed by a variety of testing experts who may join test implementations for longer or shorter periods of time. The decision to develop the IPv6 test suite based on a TTCN-3 framework was an attempt to improve the speed and quality of TTCN-3 test implementation at ETSI. Figure 14.1 illustrates how the test suite for IPv6 core aspects was implemented following our previously introduced four layered model. Most frameworks that are used today to implement TTCN-3 test suites – even in standardisation – do not directly adopt the guidelines established in the IPv6 framework in every detail. Nevertheless most of the general concepts and ideas contained in this framework are still valid and useful to illustrate some basic framework design principles.

14.4.1 Module Structure and Identifiers

In Figure 14.2 we get an idea about overall module structure of the test suite and possible invocation flows between modules in different layers. The figure also shows some

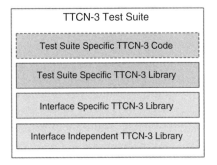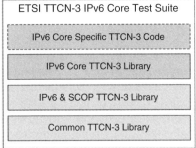

Figure 14.1 Layered test suite implementation.

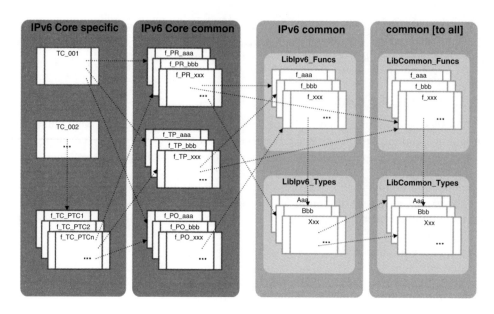

Figure 14.2 Abstracted IPv6 test suite module structure with dependencies.

identifier naming conventions that were used to make the test code more readable and traceable. In behavioural functions the framework differentiates between test case functions prefixed with 'TC_', that are functions, which are started on parallel test components, preamble functions prefixed 'PR_', postamble functions prefixed 'PO_' and test purpose functions prefixed with 'TP_', that are functions, which implement the test body. Test case as well as test case function identifiers also include a number that uniquely determines the test purpose (and requirement) for which the test has been implemented.

14.4.2 Test Case Functions

An example test case function is shown in Table 14.1. Since this test only needs one test component the test case function in this case is not a function but a testcase statement.

We see in this example the classic test phases being isolated into separate functions which are then all automatically reusable in other test specifications. This test has been specified to cover test purpose TP_COR_1432_01, which in turn has been specified based on the requirement catalogued from a RFC RQ_COR_1432. The configuration functions maps the main test component to the system component and configures the adapter, for example the Ethernet layer, for connecting the TTCN-3 test system either to an IPv6 host or router as the node under test (NUT). A generic abstract test suite default, which is defined in the second layer, is activated to ignore periodic messages that could distract from the testing objective such as neighbourhood advertisement packets. After this there follows the invocations of preamble, test body and postamble

Table 14.1   Typical IPv6 test case function implementation

```
/*
 * @desc  Tests reaction on packets that cannot be delivered not due to
 * congestion Uses test configuration 1.
 */
testcase TC_COR_1432_01()
runs on Ipv6Node
system NodeTestAdapter {

        //Variables
        var CfMessage v_cfMsg;

        //Configuration
        f_cfCore01Up(v_cfMsg);
        activate(a_tn_nut(v_cfMsg, v_cfMsg.paramsRt01, v_cfMsg.paramsIut));

        //Preamble
        f_PR_tn_nut(v_cfMsg, e_cleanGla, v_cfMsg.paramsRt01,
                    v_cfMsg.paramsIut);

        // test body
        f_TP_generateDestinationUnreachable_nonCongestion(
                        v_cfMsg.paramsRt01,
                        v_cfMsg.paramsIut,
                        v_cfMsg );

        // postamble
        f_PO_tn_nut(v_cfMsg, c_cleanGla, v_cfMsg.paramsRt01,
                    v_cfMsg.paramsIut);deactivate;

        f_cfPtcDown();

} // end TC_COR_1432_01
```

functions. The need for configuration parameters in these functions is IPv6 specific. IPv6 packets require some lower layer address information to be able to form correct packets. Finally the test configuration is torn down to bring the test system gracefully back to its initial state.

14.4.3   Test Purpose Functions

An example of an IPv6 test purpose function is shown in Table 14.2.

In this test purpose function the SUT adapter is first configured to send an IPv6 packet using an unreachable source address in the lower layer, then a generic Ipv6 library function f_getDestinationUnreachableAfterEchoReq() is configured via the parameters to fulfil the requirements set in the test purpose to send an IPv6 packet with an unreachable source address and receive in return an error, and then reconfigures the SUT adapter to its original settings. As the last step, the function attempts to

Table 14.2 IPv6 test purpose function

```
/*
 * @desc  Sends an Echo Request with a destination address to which
 *    the packet cannot be delivered due to reasons other than congestion.
 *    Checks that IUT sends a Destination Unreachable message.
 * @param p_paramsHs01 Address Information of Hs01
 * @param p_paramsIut Address Information of Node Under Test
 * @param p_cfMsg Configuration message for Test Adapter
 */
function f_TP_generateDestinationUnreachable_nonCongestion(
        template Ipv6NodeParams p_paramsHs01, template Ipv6NodeParams
        p_paramsIut, CfMessage p_cfMsg)
runs on Ipv6Node {

        var FncRetCode v_ret;
        var Ipv6Address v_origLla;
        // Create new configuration to send the packet to a broadcast MAC
        // address
        v_origLla := p_cfMsg.paramsIut.lla;
        p_cfMsg.paramsIut.lla := PX_UNICAST_UNREACHABLE_IUT;
        cfPort.send ( p_cfMsg );

        v_ret := f_getDestinationUnreachableAfterEchoReq (
                m_echoRequest_noExtHdr_noData (
                        p_paramsHs01.gla,
                        PX_UNICAST_UNREACHABLE_IUT,
                        c_defId,
                        c_defSeqNo ),
                mw_destUnreachable_code (
                        p_paramsIut.gla,
                        p_paramsHs01.gla,
                        ? ) );

        p_cfMsg.paramsIut.lla := v_origLla;
        cfPort.send ( p_cfMsg );

        f_selfOrClientSyncAndVerdict(c_tbDone, v_ret);

} // end f_TP_generateDestinationUnreachable_nonCongestion
```

synchronise with other potential test components by invoking the generally reusable function f_selfOrClientSyncAndVerdict() . Since there is only one test component in the test the call has essentially no effect. Interesting to note also that it is in this function that the actual test component verdict is set based on the current v_ret value.

Although the IPv6 framework recommends this isolation of test bodies into dedicated functions is performed, we suggest instead integrating test body code directly into the test case functions. In practice, the potential for reuse of test purpose functions in our experience is quite low and this isolation negatively affects the readability of a test since the relation of the most essential part of a test – the test body – with pre- and postamble is lost in this isolation.

Table 14.3 IPv6 library function implementation

```
/*
 * @desc   This sends an ICMPv6 echo request from an IPv6 node to any NUT,
 *         and waits for a Destination Unreachable message for a fixed amount
 *         of time.
 * @remark  Time limit is defined by module parameter PX_TAC (see comp type)
 * @param  p_echoRequest Template of the Echo Request to be sent
 * @param  p_destUnreachable Template of expected Packet Too Big message
 * @return  execution status
 */
function f_getDestinationUnreachableAfterEchoReq (
            template EchoRequest p_echoRequest,
            template DestinationUnreachable p_destUnreachable )
runs on LibIpv6Node
return FncRetCode {

      var FncRetCode v_ret;

      v_ret := f_sendEchoRequest ( p_echoRequest );

      if ( v_ret != e_success ) {return v_ret;}
      tc_ac.start;
      alt {
            []     ipPort.receive ( p_destUnreachable )  {
                   tc_ac.stop;
                   return e_success; }
            []     tc_ac.timeout{
                   return e_timeout; }
      } // end alt
} // end f_getDestinationUnreachableAfterEchoReq
```

14.4.4 Protocol Library Functions

Our final example in Table 14.3 shows an implementation of a protocol library function. This TTCN-3 function is only specified based on constructs defined in the library, which include also the function f_sendEchoRequest(). Interesting to observe here is that a variable is used to manage the verdict instead of the built-in verdict function. As we explained earlier there is also alternative approaches to handle negative verdict setting via a dedicated parameterised verdict setting function. Also the component type used in the runs on clause of this function LibIpv6Node is different from the one used in the test purpose function that invokes this function and is shown in Table 14.2. This invocation is possible because LibIpv6Node and Ipv6Node component types have been defined type compatible, that is the LibIpv6Node definition body is a subset of the Ipv6Node definition body.

14.5 Summary

This brings us to the end of this advanced chapter on TTCN-3 frameworks. We have focused in this chapter mainly on behavioural aspects to illustrate the separation of concern throughout different layers of implementation. The reader is however encouraged to study the IPv6 framework as well as other frameworks to get more detailed information on implementation guidelines, structuring and documentation of frameworks.

The main incentive for using frameworks is to isolate implementation details in a test suite implementation and to allow test case implementation to focus more on the functionality to be tested. In addition, frameworks are directly related to use of reuse. This helps to reduce time to maintain and update test suites in case of changes to interface specifications. To work with frameworks successfully and effectively in an industrial setting usually requires a central support function to be organised in a team or a company to develop, maintain and extend such frameworks, for example an organisational entity that manages generic framework development.

15

Advice and Examples

Throughout this book, we have given numerous examples to introduce the concepts of TTCN-3. But we did not say anything specific on how to write TTCN-3 in a reasonable way. However, in large projects it is important to have the TTCN-3 code well written. Therefore, in this chapter we will discuss the importance of using a TTCN-3 style guide. We will also give some further examples. These examples range from small but nevertheless useful functions to complete examples for recurring problems.

15.1 TTCN-3 Style Guide

Because TTCN-3 can in principle be considered as a programming language, all the usual issues of how to write programs have to be considered. In this section, we will show which issues should be considered in a style guide for TTCN-3 and we will explain some of the rules we followed when writing the examples in this book.

15.1.1 Motivation

To know what actually has been tested by a test case, one has to be able to *understand* it. Typically, the system under test (SUT) evolves over time and the corresponding test suite also has to evolve. Therefore, *maintainability* of the test suite is quite important. A third important issue is raised by the fact that such test suites are quite often developed by subcontractors. This is then done in collaboration with the developer of the SUT. Therefore, the test suite must be written such that it can be split into parts that each company can then reasonably work on. Such a separation into parts can be achieved by *modularisation* of the test suite.

A consistent and understandable style is always required for a test suite written in TTCN-3 as in any other programming language. Such stylistic issues include, for example how identifiers are written and how the code is laid out. Since the TTCN-3 core language

An Introduction to TTCN-3, Second Edition.
Colin Willcock, Thomas Deiß, Stephan Tobies, Stefan Keil, Federico Engler and Stephan Schulz.
© 2011 John Wiley & Sons, Ltd. Published 2011 by John Wiley & Sons, Ltd.

looks similar to C, it is relatively easy to adopt a similar layout for programs in TTCN-3. More specific issues regarding identifiers are as follows:

- **Syllables:** How should identifiers consisting of several syllables be written? Are the syllables separated by an underscore – `a_long_identifier` – or are they concatenated without separation – `aLongIdentifier`?
- **Abbreviations:** Are long identifiers abbreviated? If so, then identifiers should be abbreviated consistently. For example, `confirm` could be abbreviated to either `cnf` or `conf`.
- **Prefixes and suffixes:** How should identifiers be named when they are used in TTCN-3 operations that can be applied to different types? The `stop` operation, for example can be applied to timers, components and ports. To increase readability, it can make sense to extend identifiers by prefixes or suffixes to indicate their type.

These are just examples of what can be covered in a style guide regarding syntactic issues. The bottom line is that the names of identifiers should be meaningful.

Another important question is how a TTCN-3 test suite should be split into several modules. The modularisation has an impact on readability and maintainability. In addition, good modularisation also allows separating definitions that can then be reused in other test suites. Thus, the amount of reused code can be increased and development of test suites can become more effective. For example, all the altsteps and functions that are used to define test cases could be defined in one or more modules, whereas the test cases themselves could be defined in their own module(s). This modularisation enforces a clear interface among the building blocks – altsteps and functions – and the blocks built – the test cases. We will offer some more advice on this topic in Section 15.2.

Finally, a style guide can describe good practice in using TTCN-3. Behaviour can be described in TTCN-3 in different ways and a style guide can give advice, which way to use in a specific situation. For example, it can be clarified, which behaviour should be defined in a function, and which in an altstep. Some TTCN-3 constructs are difficult to use, and a style guide can give advice on using them – or indeed give advice on how to avoid using such constructs at all.

15.1.2 Examples

In this section, we present an example of a stylistic issue: which prefixes are used to distinguish identifiers of different kinds. Then we will give some examples of what we consider to be good TTCN-3 practice. Note that these rules do not constitute a complete style guide. Developing a style guide for a larger organisation or even for a larger test suite can be a major effort, in which a lot of issues need to be considered. To stay within the scope of this book, we restrict ourselves to presenting some examples.

- **Prefixes:** Several operations in TTCN-3 can be applied to operands of a different type, for example the `stop` operation can be applied to timers, components and ports. Similarly, a stand-alone altstep looks exactly like a function call. To enhance readability, we suggest using prefixes for identifiers. In Table 15.1, the prefixes we used in this book are shown. We did not use prefixes for type identifiers as we preferred to use the convention that type identifiers start with an uppercase letter.

Table 15.1 Prefixes of identifiers

Prefix	TTCN-3 construct	Comment
⟨none⟩	Types	Type identifiers are written without prefix, instead the first letter is written in upper case. This is consistent with type definitions written in ASN.1, which can be imported directly to TTCN-3
a_	Templates	
alt_	Altstep	The same prefix is used for altsteps independent of whether they are used as defaults or not
c_	Constants	
e_	Enumeration elements	
f_	Functions	For functions defining behaviour, purely value computing functions are written without prefix
p_	Parameters	Formal parameters of test cases, functions, altsteps, templates
pt_	Ports	
tc_	Test cases	
t_	Timers	
v_	Variables	

- **Limitation of value ranges:** In practice, almost all interfaces of an SUT allow a finite range of values to be exchanged across them. For example, `integer` values exchanged across an interface have to be within a lower and an upper bound or character strings have to be of limited length. We suggest expressing these restrictions in the type definitions by subtyping and not leaving the TTCN-3 types unrestricted. This means that the interface of the SUT should be exactly defined so that values out of range cannot be sent at all to the SUT, and neither can values out of range be received. To test whether the SUT sends messages with values out of range and how it reacts if such – invalid – messages are sent to it, further types with extended ranges can be defined and used. In this case, it is still clear which are allowed messages and which ones are not.

- **Goto:** Although TTCN-3 has `label` and `goto` statements, we discourage their usage. This is not just due to Dijkstra's famous paper [46] on this issue. Loops can be built with the `for` and `while` operations, `alt` statements can be re-evaluated by using the `repeat` operations, and more specific control flows can be described by function calls. This means there is simply no need to use `goto`.

- **Preamble, testbody, postamble:** TTCN-2 has been used often in conjunction with a methodology for *conformance testing* [ISO9646]. According to this methodology, test cases should be split into three parts: a *preamble* preparing the test system and the SUT for the actual test, a *testbody* describing the actual test and a *postamble* returning the SUT to a well-defined state. We suggest using this distinction in TTCN-3 as well, even when other approaches to testing than conformance testing are used. There are two main advantages of doing so: first, typically, several test cases have the same pre- and postamble. If these are defined as functions, then they can be easily called within test cases and the amount of duplicated code is reduced. Second, this distinction highlights the different parts in a test case and thereby helps to understand and maintain a test suite.

- **One role per test component:** As a last example, we suggest that one test component should only take one role within a test case. For example, one test component should only be connected to a single interface or a single set of closely related interfaces of the SUT. In the Domain Name System (DNS) example in Chapter 2, there was one test component taking the role of the local client, one taking the role of a root name server and one taking the role of a remote name server. Even the main test component (MTC) had just a single role: controlling the parallel test components (PTCs). It did not have a role in testing towards the SUT. This allows a single aspect to be focused on when defining the behaviour of a single test component.

In this section, we have shown several examples of good practice when writing TTCN-3. These examples are by no means a complete style guide, which would be beyond the scope of this book and most probably would not match existing practices in organisations. Nevertheless, these examples show that a style guide is helpful when writing large test suites and that such a style guide covers a wide range from purely syntactical issues to best practices of TTCN-3.

It is common practice in software projects to define a style guide to increase understandability and maintainability of the developed code. Note that in test systems it might be even more important to have well-written and easy-to-maintain code: if it is unclear what a test case actually does, then it is not much use that a test case has passed. And if it is difficult to adapt a test suite to even small changes in functionality; then a test system will be more of a burden than an aid and its further development will soon be abandoned.

15.2   Suggestions for Modularisation

In Chapter 7, we have argued that there are many good reasons for separating your test suite into separate modules, which is one of the most fundamental concepts of software engineering. The TTCN-3 core language standard itself neither mandates nor suggests any guidelines on how a test suite should be modularised. The main reason is that approaches to modularisation may depend on the specific application area of TTCN-3. In this section we try to propose some possible guidelines for test suite modularisation, which is mainly targeted towards the domain of protocol testing. The result of these guidelines is illustrated in Table 15.2. Note that although we show all module definitions in this table, we encourage saving each module definition in a separate file.

The most obvious definitions to isolate in their own modules are protocol type definitions for the interface used by the test suite. Naturally, there should be one module per protocol or interface. An example of such a module is `ProtocolTypes`. Another intuitive part to separate into its own module is the test execution control, that is the TTCN-3 control part. This module simply imports all modules that define test cases and describes in which order they are to be executed. This module is called `TestControl` in our example. Type definitions are required to build up test configurations, for example TTCN-3 component type definitions, the abstract test system interface (TSI), as well as port type definitions, are a third good candidate for a separate module. These definitions must be used in `runs on` or `system` clauses of any test case implementation and should

Table 15.2 Example modularisation of a protocol testing test suite

```
// File: ProtocolTypes.ttcn

module ProtocolTypes {
  // Example: TTCN-3 constants & type definitions for DNS message structure
}

// File: TestSystem.ttcn

module TestSystem {
  import from ProtocolTypes { type all };

  // Example: TTCN-3 definitions of DNS client & TSI ( i.e., system )
  //          component type and port types for DNS interface, possibly
  //          types for component variables, etc.
}

// File: TestControl.ttcn

module TestControl {
  import from TestCases { testcase all };

  // Example: Specification of test case execution sequence;
  //          This may also import more test case modules if there is more;
  //          Test case selection may also be driven by module parameters
}

// File: TestSuiteMPs.ttcn

module TestSuiteMPs {
  // Example : Module parameter definitions needed to configure a test suite
  //           for a specific test execution, e.g., the SUT IP address
}

// File: TestCases.ttcn

module TestCases {
  import from ProtocolTypes all;
  import from ProtocolTemplates all;
  import from TestSystem all;
  import from TestSuiteMPs { modulepar all };
  import from Functions all;

  // Example : Specification of DNS test cases; here "test case" means
  //           definition of MTC behavior ( = test case statement ) but also
  //           if applicable PTC ( = function definitions ) behavior;
  //           each test case must at least establish the required test
  //           configuration and may then invoke other "Functions" to drive
  //           the interaction with the SUT
}
```

Table 15.2 (*continued*)

```
// File: ProtocolTemplates.ttcn

module ProtocolTemplates {
  import from ProtocolTypes all;
  // Example: TTCN-3 template definitions for DNS messages
}


// File: Functions.ttcn

module Functions {
  import from ProtocolTypes all;
  import from ProtocolTemplates all;
  import from TestSystem all;
  import from TestSuiteMPs { modulepar all };

  // Example: TTCN-3 function and altstep definitions for common behavior,
  //          e.g, pre- & postambles like configuring the transport of DNS
  //          messages to the SUT, or handling of interface shutdown in case
  //          of problems in test case executions
}
```

therefore be easy to find in a test suite. Because port type definitions require message types a protocol definition module needs to be imported into this module. This module is called `TestSystem` in our example.

The modularisation of the actual test behaviour is without doubt the biggest challenge. One approach is to use one or more modules for storing test-case *specific* behaviour, that is the TTCN-3 test case statement as well as the functions that are first started on PTCs. The criteria for assigning such behaviour to a given 'test case module' could be, for example the SUT features that it tests. An example of such a module is `TestCases`. Secondly, behaviour that can be reused by multiple test cases should be separated into other modules, that is message interchanges like preambles or postambles and also algorithms. An example of such a module is `Functions`. Finally, separate modules for constant value definitions like module parameters and template definitions simplify the reuse of these definitions across multiple test case implementations. Examples of such modules are `TestSuiteMPs` and `ProtocolTemplates`.

In the long run, the challenge in modularisation is to keep the number of modules and module definition sizes in proper balance. Too many module definitions may also make the management of a test suite implementation harder. But even the presented modularisation leaves room for additional modules. A good example could be a collection of test component synchronisation routines, which are needed for test cases with multiple test components and will be discussed in Section 15.5.

15.3   Template Specification for Complex Message Definitions

The decomposition and parameterisation of templates could be considered one of the most difficult topics to master for newcomers to the language. Real test suites can have hundreds or even thousands of TTCN-3 test cases, which in turn can contain a substantial

number of message exchanges with complex message content. However, when starting a
test suite implementation, it may be quite hard to select how to specify template definitions
in order to gain the maximum amount of reuse from them.

In general, template definitions should be decomposed once information starts to be
repeated within as well as across multiple template definitions. At the same time you want
to avoid the two extremes, which are to specify one single generic template per type definition or to specify a complete message in one single template definition, that is to define
many very specific templates. One possibility is to choose major information elements
within a message as a basis for template decomposition. The term *information element*
could be loosely defined as a 'fairly self-contained block of information within a message'.
These information elements would then be completely specified within one template definition. Some aspects of their information content could then be parameterised to increase
their reuse. A separate template definition for the complete message would then compose
the message from these information element templates.

15.3.1   Example Implementation of a SIP Message Interchange

In Chapter 8, we introduced the Session Initiation Protocol (SIP) protocol with an
example message exchange, that is shown again in Figure 15.1. In this message, one SIP
user, Alice, tries to establish a SIP session with another SIP user, Bob. In the following
sections, we will show how one could implement a TTCN-3 function, which acts as
Bob and correctly replies with a 200 OK response to the INVITE from Alice. The main
focus in this exercise will be the definition, decomposition and parameterisation of the
templates used by this function.

15.3.2   A SIP Type Definition

Before we can start implementing a TTCN-3 function, which sends and receives a message, we need to specify templates for the messages that are to be sent and received.

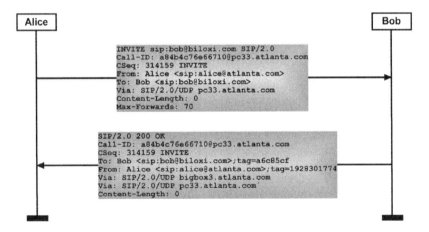

Figure 15.1   Example SIP message exchange.

However, in order to specify such message instances, we must first have a type structure that specifies the format of such messages, that is a protocol type definition. Unfortunately, the SIP standard [26] does not supply us directly with an abstract syntax definition for SIP messages. Let us assume that we have *derived* such a definition shown in Table 15.3, for example by using the approach described in [35].

Our example SIP protocol definition separates messages into requests and responses. Each message carries as its major information elements a first line specific to the message type as well as a list of headers. The headers can occur in both kinds of messages. Another important class of information elements is addressing types for SIP. These are used by a number of header types as well as the `RequestLine` type. Although this example is still not a very complex protocol type definition, it is sufficient to explain the structuring of the templates.

15.3.3 Specification of the Expected SIP Request

Our first step is to define the INVITE message, that is expected to be received. Here, we have the possibility of making use of the powerful TTCN-3 matching expressions. Thinking a bit ahead towards other message interchanges in our test suite, we specify our message template using template parameters so that it can be used for receiving (and also sending) any kind of `SipRequest`, as shown in Table 15.4. Remember that template parameters do not only enable us to pass in matching expressions, but also references to other templates, for example for first line and header information elements.

When a message is expected, its template definition should in general only focus on the aspects highlighted in the test purpose. In our case, we only really care that the expected message is an INVITE for Bob. We ensure this by checking these values in the `a_inviteBob` and `a_bobSipUri` templates. The remaining information is accepted as long as the received SIP request follows the SIP message syntax. The rational for separating the `Uri` from the `RequestLine` template definition is that the Uniform Resource Identifier (URI) constitutes a major information element, which appears in many parts of a message. This means that the `a_bobSipUri` template is likely to be useful in future message template definitions.

15.3.4 Specification of the 200 OK Response

The specification of our second message requires the definition of a single value template since it is going to be used in a TTCN-3 `send` operation. In our definition of this complete message value, we again decompose the message value along the lines of the major information elements. Notice that in Table 15.5 we can already reference or reuse our previous `a_bobSipUri` template in the `a_bobToTagHdr` template.

It turns out that within a SIP message exchange, a lot of information must be returned, for example caller identification, sequence number, the caller's 'Via' header and so on. This issue has been handled by using template parameterisation in the `a_bobInviteRespHdrs` template, which allows us to easily assign such information at run time. Note that some parameter values may have to traverse multiple template definitions until they get assigned, for example `p_fromNameAddr`.

Table 15.3 An example structured SIP TTCN-3 protocol type definition

```
module SipTypes {
  group specialTypes {
    type component SipUserAgent { SipPort pt_sip }
    type port SipPort message { inout SipRequest, SipResponse }
  } // end group specialTypes

  group msgTypes {
    type record SipRequest {
      RequestLine requestLine,
      SipHeaders  reqHdrs,
      charstring  messageBody optional
    }

    type record SipResponse {
      StatusLine statusLine,
      SipHeaders resHdrs,
      charstring messageBody optional
    }
  } // end group msgTypes

  group firstLineTypes {
    type record RequestLine {
      charstring method,
      Uri        requestUri,
      SipVersion version
    }

    type record StatusLine {
      SipVersion     version,
      SipStatusCode  code,
      charstring     reasonPhrase
    }
  } // end group firstLineTypes

  group headerTypes {
    type set SipHeaders {
      charstring          callIdHdr,
      Cseq                cSeqHdr,
      From                fromHdr,
      To                  toHdr,
      Via_List            viaHdr,
      UInt                contentLengthHdr optional,
      OtherHeader_List otherHdrs optional
    }
    type record  Cseq {
      UInt       seqNo ,
      charstring method
    }
    type record From {
      AddrField addrField,
      GenericParam_List fromParams optional
    }
```

Table 15.3 (*continued*)

```
    type record To {
      AddrField addrField,
      GenericParam_List toParams optional
    }
    type record ( 1..infinity ) of Via Via_List;
    type record Via {
      charstring        sentProtocol,
      HostPort          sentBy,
      GenericParam_List viaParams optional
    }
    type set ( 1..inifinity ) of OtherHeader OtherHeader_List;
    type record OtherHeader {
      charstring hdrName,
      charstring hdrValue
    }
} // end group headerTypes

group addressingTypes {
    type union AddrField { NameAddr nameAddr, Uri addrSpec }
    type record NameAddr { charstring displayName optional, Uri addrSpec }
    type union Uri { SipUri sip, SipUri sips, charstring absoluteUri }
    type record SipUri {
      UserInfo           userInfo optional,
      HostPort           hostPort,
      GenericParam_List  urlParams optional,
      GenericParam_List  urlHeaders optional
    }
    type record UserInfo { charstring user, charstring passwd optional }
    type record HostPort { Host host, UShort portField optional }
    type union Host {
      charstring ipv4,
      charstring ipv6,
      charstring hostName
    }
} // end group addressingTypes

group miscTypes {
    type set ( 1..inifinity ) of GenericParam GenericParam_List;
    type record GenericParam { charstring id, charstring pValue optional }
    type charstring SipVersion ( "SIP/2.0" );
    type integer SipStatusCode ( 100..606 );
    type integer UInt ( 0..infinity );
    type integer UShort ( 0..65535 );
} // end group miscTypes
} // end module
```

Table 15.4 Definition and example use of SIP INVITE templates

```
template SipRequest a_sipReq( template RequestLine p_reqLine,
                              template SipHeaders  p_sipHdrs,
                              template charstring  p_msgBody ):= {
  requestLine := p_reqLine,
  reqHdrs     := p_sipHdrs,
  messageBody := p_msgBody
}

template RequestLine a_inviteBob := {
    method     := "INVITE",
    requestUri := a_bobSipUri, // 1st reference to template!
    version    := "SIP/2.0"
}

template Uri a_bobSipUri  := {
  sip := {
    userInfo  := { user := "bob", passwd := omit },
    hostPort  := { host := { hostName := "biloxi.com"}, portField
              := omit },
    urlParams := omit,
    urlHeaders := omit
  }
}

function f_waitForBobInvite () runs on SipUserAgent {
  timer t;
  t.start( 10.0 );
  alt {
    [] pt_sip.receive( a_sipReq( a_inviteBob, ?, omit ) ) { t.stop; }
    [] pt_sip.receive { log( "This was not an INVITE for Bob!" ); }
    [] t.timeout      { log(."Where is Alice?" );}
  } // end alt
} // end function f_waitForBobInvite
```

Finally, Table 15.6 shows the function, which would model Bob as described previously in Figure 15.1. Here, TTCN-3 value redirection is used to capture the received information in the `v_sipReq` variable, which is then used to configure a correct SIP response.

15.3.5 About the Benefits of Smart Template Definitions

The main benefit of smart template decomposition and parameterisation is the ability to reuse template definitions in other message exchanges within a test suite and to reduce the test suite complexity and size. One example is the expansion of our SIP test suite to also handle the shut down of a session, which is done by sending a SIP BYE message. Here, only two new template definitions have to be introduced to handle this additional message exchange, as shown in Table 15.7.

Table 15.5   A SIP 200 OK response definition

```
template SipResponse a_sipRespNoMsgBody( template RequestLine p_statusLine,
                                         template SipHeaders   p_sipHdrs )
                                         := {
  statusLine   := p_statusLine,
  resHdrs      := p_sipHdrs,
  messageBody := omit
}


template StatusLine a_200ok := {
    version      := "SIP/2.0",
    code         := 200,
    reasonPhrase := "OK"
}


template SipHeaders a_bobInviteRespHdrs ( in Cseq       p_cseq,
                                          in UInt       p_seqNo,
                                          in NameAddr   p_fromNameAddr,
                                          in Via        p_sdrViaHdr ) := {
  callIdHdr        := p_callId,
  cSeqHdr          := p_cseq,
  toHdr            := a_bobToTagHdr,
  fromHdr          := a_fromTagHdr( p_fromNameAddr ),
  contentLengthHdr := 0,
  viaHdrs          := { p_sdrViaHdr , a_bobProxyViaHdr }
  otherHdrs        := omit
}


template To a_bobToTagHdr   := {
  addrField := {
    nameAddr := {
      displayName := "Bob",
addrSpec := a_bobSipUri // 2nd reference to template!
    }
  },
  toParams := {{ "tag", "a6c85cf" }}
}


template From a_fromTagHdr( in NameAddr p_nameAddr ) := {
  addrField  := { nameAddr := p_nameAddr },
  fromParams := {{ "tag", "1928301774" }}
}


template Via a_bobProxyViaHdr := {
    sentProtocol := "SIP/2.0/UDP",
    sentBy       := {
      host := { hostName := "bigbox3.atlanta.com" },
      portField := omit
    },
    viaParams := omit
}
```

Table 15.6 The INVITE message exchange

```
function f_bob () runs on SipUserAgent {
  timer t;
  var SipHeaders v_sipHdrs;
  t.start( 10.0 );
  alt {
    [] pt_sip.receive( a_sipReq( a_inviteBob,?,omit ) )
         -> value( v_sipHdrs := SipRequest.reqHdrs) {
       t.stop;
    }
    [] pt_sip.receive {
       log( "This was not an INVITE for Bob!" );
       return;
    }
    [] t.timeout { log( "Where is Alice?" ); return; }
  } // end alt
    pt_sip.send( a_sipRespNoMsgBody(
              a_200ok,
              a_bobInviteRespHdrs ( v_sipHdrs.callIdHdr,
                                    v_sipHdrs.cseqHdr,
                                    v_sipHdrs.fromHdr.addrspec.nameAddr,
                                    v_sipHdrs.viaHdrs[0] )
              ) );
  return;
} // end function f_bob
```

15.4   Useful Behaviour

When writing test suites, we often experience recurring behaviour. Although this behaviour is often quite simple, it is nevertheless useful. The corresponding definitions could be defined as *useful functions* in a single TTCN-3 module and imported into a test suite whenever needed. In this section, we will present some examples of such useful functions.

15.4.1   Convert Conditions to Verdicts

Often, it can be expressed in the templates of receiving operations whether a certain response of the SUT is the expected one or not. However, sometimes this is not possible, for example when fields of subsequently received messages have to satisfy a specific condition. In this case, it is reasonable to evaluate an appropriate Boolean expression and set the verdict according to the outcome of this evaluation. To avoid cluttering the TTCN-3 code with `if` statements, the function shown in Table 15.8 is useful.

15.4.2   Unexpected Messages

When waiting for a reaction from the SUT, one typically describes which messages are expected on which port and what should happen then. However, it can happen that messages on another port are received that are either not expected at this point or even

Table 15.7   A SIP INVITE and BYE message exchange

```
template RequestLine a_byeBob modifies a_inviteBob := {
    method := "BYE"
}

template SipHeaders a_bobByeRespHdrs ( in Cseq        p_cseq,
                                       in UInt        p_seqNo,
                                       in NameAddr    p_fromNameAddr,
                                       in Via         p_sdrViaHdr )
modifies a_bobInviteRespHdrs:= {
  viaHdrs := { p_sdrViaHdr } // now do not return Bobs proxy
}

function f_bobPlusBye() runs on SipUserAgent {
  timer t;
  var SipHeaders v_sipHdrs;
  f_bob(); // responds to SIP INVITE
  t.start( 10.0 );
  alt {
    [] pt_sip.receive( a_sipReq( a_byeBob,?,omit ) )
       -> value v_sipHdrs := SipRequest.reqHdrs {
         t.stop;
       }
    [] pt_sip.receive { log( "This was not a BYE for Bob!" ); return; }
    [] t.timeout { log( "Where is Alice?" ); return; }
  } // end alt
  pt_sip.send( a_sipRespNoMsgBody(
                a_200ok,
                a_bobByeRespHdrs ( v_sipHdrs.callIdHdr,
                                   v_sipHdrs.cseqHdr,
                                   v_sipHdrs.fromHdr.addrspec.nameAddr,
                                   v_sipHdrs.viaHdrs[0] )
              ) );
  return;
} // end function f_bobPlusBye
```

Table 15.8   Setting the verdict according to a Boolean condition

```
function f_assert ( in boolean p_cond ) {
  if ( p_cond ) {
    setverdict( pass )
  } else {
    setverdict( fail )
  }
};
```

Table 15.9   Receive unexpected messages

```
altstep alt_catchAll() {
  [] any port.receive {                          // 'catch all'
     log( "unexpected message received" );
     setverdict( fail );
     mtc.stop;                                   // force shutdown
     }
}
```

not expected at all. In such a situation the test case should fail. These messages can be detected quite easily by activating the altstep in Table 15.9 as a default. Using this default, only the expected cases have to be described in the test cases and functions.

Whether the verdict should be set to fail and the test case terminates by stopping the MTC depends on the specific test suite implementation.

15.4.3   Waiting

Minimal separation in time between two subsequent events in a behaviour description can be expressed easily by starting a timer and waiting for its expiration. By defining a function with its own timer, a separate declaration of the timer can be avoided wherever minimal separation should be expressed. Such a function can be defined as shown in Table 15.10.

Note that in the timeout statement the currently active defaults will be considered. If a message is received while the timer has not yet expired and there is a default to handle this message, then this default will be executed. Therefore, this function only roughly expresses minimal separation. Because the currently activated defaults cannot be suspended and resumed in a general way in TTCN-3, another approach has to be taken to avoid the evaluation of defaults. In this other approach, an else branch with a repeat statement is added after the timeout statement. As the else branch can always be taken, no default will be executed. However, this approach also has its price; this is actually a busy waiting loop. Nevertheless, the corresponding function is shown in Table 15.11.

This busy waiting can be avoided in certain situations. In general, it is not possible to retrieve which defaults are currently active and to suspend them. But if there are only a limited number of defaults in a test suite that can be active, then it is possible to suspend these defaults while waiting as shown in Table 15.12. In order to improve the maintainability of a test case we suggest keeping the number of defaults very small. It

Table 15.10   Wait for some time

```
function f_wait ( in float p_duration ) {
  timer t;
  t.start( p_duration );
  t.timeout;
}
```

Table 15.11   Wait without defaults

```
function f_waitBusy ( in float p_duration ) {
  timer t;
  t.start( p_duration );
  alt {
    [] t.timeout {}
    [else] {repeat}
  }
}
```

Table 15.12   Waiting without being interrupted

```
type component CatchAllComp {
  var default v_catchAllDef := null;
};

function f_waitUninterrupted( in float p_duration ) runs on CatchAllComp {
  timer t;
  var boolean v_altWasActive := false;
  if ( v_catchAllDef != null ) {              // move default out of the way
    deactivate( v_catchAllDef );
    v_altWasActive := true;
  }

  t.start( p_duration );
  t.timeout;
  if ( v_altWasActive ) {                     // re-activate default
    v_catchAllDef := activate( alt_catchAll() );
  }
}
```

is even reasonable to activate only the altstep alt_catchAll shown in Table 15.9. In Table 15.12, a component variable v_catchAllDef is defined to hold the reference if this altstep is activated. The function f_waitUninterrupted checks first whether the default has been activated and, if so, it is deactivated and this fact is stored into the variable v_altWasActive. Then, after waiting, the altstep can be activated again.

Note that although the runs on clause of this altstep refers to the component type CatchAllComp, this code can be executed on any component that has a variable to store default references and which is named v_catchAllDef.

15.4.4 Successful Altstep

Usually, it depends on an operation such as receive or timeout whether an alternative in an alt statement or an altstep will be chosen. The Boolean guard expression is seen as an additional means to choose which alternative can be chosen. However, in some cases it is useful to express that an alternative is chosen depending only on the Boolean guard expression. For syntactical reasons, there has to be some operation between the guard and

Table 15.13 Successful altstep

```
altstep alt_else () {
  [else] {}
}


function f_receive ( in integer p_n,
                     in float   p_duration ) {
  var integer v_received := 0;
  timer t_duration;

  t_duration.start( p_duration );
  alt {
    [v_received <  p_n] any port.receive {
      v_received := v_received + 1;
      repeat;
      }
    [v_received >= p_n] alt_else() {
      setverdict( pass );
      }
    [] t_duration.timeout {
      setverdict( fail );
      }
    }
  }
}
```

the statement list of an alternative. If this statement is the call to an altstep, that is always successful, then it depends solely on the guard whether this alternative will be taken. In the example in Table 15.13, at first the altstep `alt_else` is defined, which will always execute successfully. Thereafter, the usage of this altstep is shown in a function that tries to receive a certain amount of messages within a given period of time. The altstep `alt_else` is used in cases where a sufficient amount of messages has been received.

15.4.5 Additional String Conversion Functions

Although the TTCN-3 core language lacks some native string conversion functions, it is possible to write your own functions for this purpose. Table 15.14 shows an implementation of converting a `charstring` to a `universal charstring` value, and vice versa. Note that here not all `universal charstring` values have a `charstring` value counterpart, that is if they contain a non-ASCII character.

Similarly, it is also possible to define a function for converting a `charstring` into a `float` value as shown in Table 15.15. If you study this definition of `str2float` closely, you will notice that there are classes of (illegal) arguments for which it will not yield the correct results. For example, `str2float("2.illegal argument")` yields `2.0`. It would be a good exercise for you to try and make this function first check its arguments for the expected format. However, for us it would exceed the space that we want to spend on this example.

Table 15.14   charstring and universal charstring conversion functions

```
function f_str2unistr( in charstring p_str ) return universal charstring {
  var integer v_index;
  var integer v_length := lengthof( p_str );
  var universal charstring v_result := "";

  for ( v_index := 0; v_index < v_length; v_index := v_index + 1 ) {
    v_result := v_result & int2unichar( char2int( p_str[v_index] ) );
  }
  return v_result;
}

function f_unistr2str( in universal charstring p_ustr ) return charstring {
  var integer v_index;
  var integer v_length := lengthof( p_ustr );
  var charstring v_result := "";

  for ( v_index := 0; v_index < v_length; v_index := v_index + 1 ) {
    // prevent 'overflow' of int2char {
    if ( unichar2int( p_ustr[v_index] ) <= 127 )
      v_result := v_result & int2char( unichar2int( p_ustr[v_index] ) );
    }
    else {
      log ( "F_unistr2str: Unable to convert universal
            charstring character!" );
      return "ERROR in f_unistr2str!";
    }
  }
  return v_result;
}
```

15.4.6 Binary Addition

TTCN-3 does not include an addition operator for bitstring values. Again, it is possible to define your own function by using available binary string type operations to construct binary addition as shown in Table 15.16. The presented algorithm assumes that the bitstrings to be added are of equal length and that both start with the least significant bit at index 0.

15.5   Test Component Synchronisation

When multiple test components are used in a test case, their execution needs to be coordinated and synchronised. In our first examples of multi-component test cases in Chapter 5, synchronisation was achieved implicitly by making sure that test components wait at relevant points to receive a message either from the SUT or from another test component. In this section, we will now have a look at two general approaches to synchronise test execution with multiple test components: one based on alive test components and one using a synchronisation protocol.

Table 15.15 Conversion of a string to a `float` value

```
function f_str2float( in charstring p_str ) return float {
  var integer v_length := lengthof( p_str );
  var integer v_dotPosition := 0, v_fractionalLength;
  var boolean v_dotFound := false;
  var float v_result := 0.0;
  var float v_fractional := 0.0;

  // find the '.' in the argument
  for ( v_dotPosition := 0;
        ( v_dotPosition < v_length ) and not v_dotFound;
          v_dotPosition := v_dotPosition + 1 )
  {
    if ( p_str[v_dotPosition] == "." ) { v_dotFound:= true; }
  }

  if ( not v_dotFound ) {
    log( "f_str2float: Not a float string - returning 0.0!" );
    return v_result;
  }

  // extract integral and fractional part
  v_result := int2float( str2int( substr( p_str, 0, v_dotPosition ) ) );
  v_fractional := int2float( str2int( substr( p_str,
                                       v_dotPosition + 1,
                                       v_length - v_dotPosition - 2 )
                                     ) );

  // shift fractional part behind the '.'
  v_fractionalLength := v_length - v_dotPosition;
  while (v_fractionalLength > 0) {
    v_fractional := v_fractional / 10.0;
    v_fractionalLength := v_fractionalLength - 1;
  }

  // combine integral and shifted fractional part
  if ( v_result < 0.0 or p_str[0] == "-" ) {
    v_result := v_result - v_fractional;
  } else {
    v_result := v_result + v_fractional;
  }

  return v_result;
};
```

Table 15.16 Addition of `bitstring` values

```
function f_add4b( in bitstring p_x, in bitstring p_y ) return bitstring {
  var bitstring v_result := ''B;
  var bitstring v_carry := '0'B;
  var integer v_index;

  if ( lengthof( p_x ) != lengthof( p_y ) ) {
    return ''B;
  }

  for ( v_index := 0; v_index < lengthof( p_x ); v_index := v_index + 1 ) {
    v_result := v_result & ( p_x[v_index] xor4b p_y[v_index] xor4b v_carry );
    v_carry  :=           ( v_carry and4b ( p_x[v_index] or4b p_y[v_index] ) )
            or4b ( p_x[v_index] and4b p_y[v_index] );
  }

  if ( v_carry == '1'B ) {
    v_result := v_result & '1'B;
  }

  return v_result;
}
```

Before we explain these approaches let us first review some of the basic principles of test specification. According to [47], at least the three phases preamble, test body and postamble should be distinguished in a test case. The purposes of the three phases can be described as follows:

- **Preamble:** This section of the behaviour starts from a defined state of the SUT and brings it to a state where the actual test can take place. An example of this is setting up the connections with the SUT.
- **Test body:** This part of the behaviour is the actual test.
- **Postamble:** This part of the behaviour takes the SUT back to a defined state, such that another test case can start afterwards. An example of this is tearing down the connections with the SUT.

In the context of a multi component test case the behaviour of each test component follows these three phases. In addition, all test components should be synchronised with each other prior to the execution of the test body and the postamble, since their order of execution within the test as well as the order of responses by the SUT to different test components may not predictable. Figure 15.2 displays one example where each test component executes through its preamble, test body and postamble in a fully synchronised manner.

In this section, we will introduce two alternatives for implementing test component synchronisation in TTCN-3. As depicted in Figure 15.2, we will assume that a test uses a MTC and *n* PTCs. The names of the component types and of their execution phases, implemented by functions, will be as depicted in the figure. For the sake of simplicity,

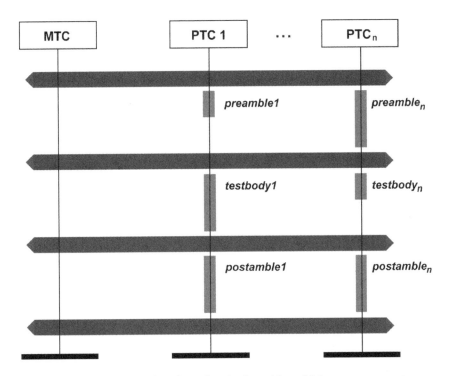

Figure 15.2   Message based synchronisation with multiple test components.

we assume that the phase functions do not have parameters and that they all have a `runs on` clause for the corresponding component type, for example `f_preamble1()` runs on `PTC1`.

15.5.1   Synchronisation with Alive Components

As we introduced previously in Chapter 12 there are two different execution modes for TTCN-3 test components. One is the regular mode which allows behaviour to be started only once on a test component instance, and an alive mode where behaviour can be started as often as desired. In this section, we show how multiple test components can be synchronised by using alive components. In this approach, the different phases of each test component behaviour are implemented in separate functions. We select the MTC to act as the coordinator of test execution which creates all PTCs and then starts the behaviour on each test component as required. A simple implementation of this approach with two PTCs is shown in Table 15.17. Figure 15.3 shows that the synchronisation in this approach is actually also achieved implicitly by messages, that is the component signalling in the TTCN-3 run time engine for start and done handling. Note that this example does not handle the case when test components get blocked in their execution. In such cases a guard timer should be used to prevent the overall test case getting blocked during execution.

Table 15.17 Example test synchronisation with alive test components

```
testcase tc_aliveSync () runs on MTC {
  var PTC1 v_ptc1;
  var PTC2 v_ptc2; //
  // set up test configuration
  v_ptc1 := PTC1.create alive;
  v_ptc2 := PTC2.create alive;
  // first  synchronisation point
  v_ptc1.start( f_preamble1() );
  v_ptc2.start( f_preamble2() );
  all component.done; // waits until all PTCs stop executing
  // second  synchronisation point
  v_ptc1.start( f_testbody1() );
  v_ptc2.start( f_testbody2() );
  all component.done;
  // third  synchronisation point
  v_ptc1.start( f_postamble1() );
  v_ptc2.start( f_postamble2() );
  all component.done;
  stop;
}
```

15.5.2 Synchronisation via a Protocol

Another approach for handling test component synchronisation is to use regular components and protocol to synchronise the execution of all test components. Remember that it is not possible in TTCN-3 to start functions more than once on test components which are not defined with the alive keyword. One way to design such a synchronisation protocol is to select the MTC to act as the synchronisation master and all other, PTCs as slaves.

We start by defining our synchronisation protocol in Section 15.5.2.1. The synchronisation code for the PTCs is explained in Section 15.5.2.2 and the code for the MTC is shown in Section 15.5.2.3.

15.5.2.1 Example Synchronisation Protocol

To be able identify more easily between different phases in test execution logs we define an enumeration type Phase as part of our synchronisation protocol. The actual message types used to synchronise the test components are PhaseStartReq and PhaseEndInd. Figure 15.4 shows an example where messages of these types have one parameter indicating the phase.

A message of type PhaseStartReq is used to request a PTC to start a specific phase. A message of type PhaseEndInd is used by each PTC to inform the MTC that it has finished a specific phase. These definitions are shown together with port and component definitions for these messages in Table 15.18. As the MTC will take the master role for the synchronisation while the PTCs act as slaves, this requires us to define different port types accordingly.

We will define some of the functions needed for synchronisation using general synchronisation specific component types. As long as the actual component types for the

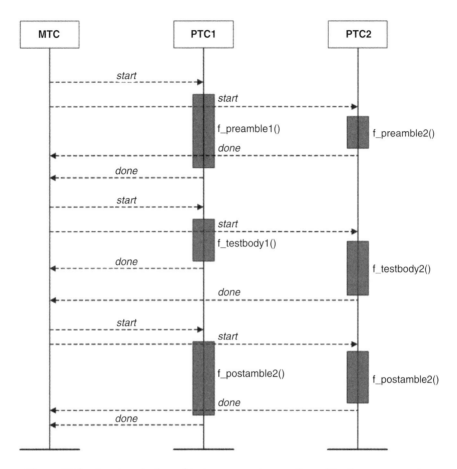

Figure 15.3 Synchronisation of test component execution with alive components.

MTC and the PTCs have corresponding ports defined, that is component types are type compatible, it is possible to execute these functions also on the specific test components.

The part of the test configuration, that is relevant for the synchronisation is shown in Figure 15.5. There can be further connections and mappings in addition to those shown in the figure.

15.5.2.2 Handling of Synchronisation by Parallel Test Components

On each of the PTCs preamble, test body and postamble are executed under control of the MTC. Before each of the corresponding functions can be executed, the PTCs have to wait for a message from the MTC. After terminating such a function, the PTCs have to signal to the MTC that they have finished a phase. The current phase is stored in the variable v_phase of the PTC. The component type SyncSlave shown in Table 15.19 defines a port pt_sync and a variable v_phase. Each of the component types of the PTCs needs to define this port and also this variable.

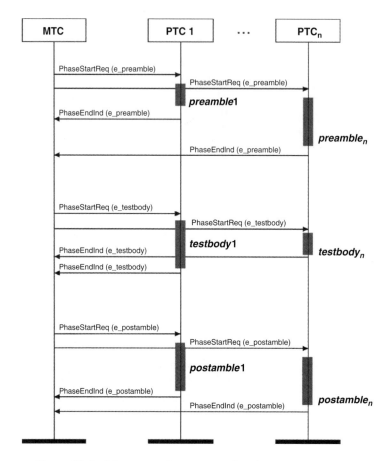

Figure 15.4   Message exchange to synchronise test components.

Table 15.18   Synchronisation message and port definitions

```
type enumerated Phase { e_preamble, e_testbody, e_postamble };
const Phase c_firstPhase := e_preamble;

type record PhaseStartReq { Phase phase };
type record PhaseEndInd   { Phase phase };

type port SyncMasterPort message {
  out PhaseStartReq;
  in  PhaseEndInd
};

type port SyncSlavePort message {
  in  PhaseStartReq;
  out PhaseEndInd
};
```

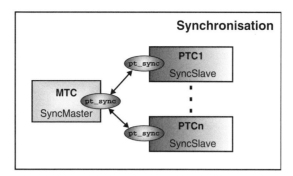

Figure 15.5   Test configuration with synchronisation components.

Table 15.19   General synchronisation component type for synchronisation slaves

```
type component SyncSlave {
  port SyncSlavePort  pt_sync;
  var  Phase          v_phase := c_firstPhase
};
```

Table 15.20   Common slave synchronisation function to wait to start execution in a given phase

```
altstep alt_awaitPhaseStartReq () runs on SyncSlave {
  [] pt_sync.receive( PhaseStartReq:{ phase := v_phase } )
     {};
  [] pt_sync.receive( PhaseStartReq:{ phase := ? } ) {
     setverdict( inconc ) // handling of out-of-sync
     }
}
```

Next, we define an altstep `alt_awaitPhaseStartReq` in Table 15.20 that waits for
a message of type `PhaseStartReq` with the corresponding phase as parameter. The first
of the alternatives matches the current phase; the second alternative matches messages
of this type, but with another phase as parameter. In the latter case, something has gone
wrong in the test case and therefore nothing more can be said about success or failure of
the test case. The local verdict is set to `inconc`.

In both cases, the message is received and the behaviour of the PTC can proceed. After
the behaviour of a phase has finished, a PTC sends a message of type `PhaseEndInd`
with the just-ended phase to the MTC. The PTC then adjusts its phase to the next one by
calling the `f_incPhase` function, which simply updates the current phase information.
Indicating the end of a phase to the MTC is done together with adjusting the phase in a
single function. This function is shown in Table 15.21.

The function calls for the phases, the altsteps and functions for synchronisation are put
together in a single function for each of the PTCs as shown in Table 15.22 for the first
PTC. For each of the other PTCs, a similar function has to be defined.

Table 15.21 Common slave synchronisation function for indicating the end of a phase to the master

```
function f_sendPhaseEndInd ( ) runs on SyncSlave {
  pt_sync.send( PhaseEndInd: { phase := v_phase } );

  v_phase := f_incPhase( v_phase ); // advance test component to next phase

  return;
}
```

Table 15.22 Example use of common slave synchronisation functions

```
function f_ptc1 () runs on PTC1 {
  alt_awaitPhaseStartReq();    // point of synchronization
  f_preamble1();
  f_sendPhaseEndInd();

  alt_awaitPhaseStartReq();    // point of synchronization
  f_testbody1();
  f_sendPhaseEndInd();

  alt_awaitPhaseStartReq();    // point of synchronization
  f_postamble1();
  f_sendPhaseEndInd();

  stop;
}
```

Please observe that the listed code does not bother with timing constraints, but such constraints can easily be added by passing timing information between the different components.

15.5.2.3 Handling of Synchronisation by Main Test Component

A single port of the MTC will be connected to one port of each of the PTCs. To keep things simple in our synchronisation protocol we then sent messages explicitly to each of the PTCs. To keep track of existing PTCs, we define a type to hold a set of component references. The definition of this set in Table 15.23 is very simplistic as we are focusing on synchronisation aspects in this example.

The elements in the set are defined to be of type SyncSlave, but it is possible to store component references to any component type that has the port pt_sync and variable v_phase defined. In the behaviour of the MTC, we add the component reference to each PTC to the set immediately after its creation. The set itself is stored in the variable v_slaveSet in the MTC as shown in the definition of SyncMaster in Table 15.24.

To invoke the execution of a given phase, the MTC sends a message of type PhaseStartReq with the indicated phase to all PTCs. In the function f_startPhase shown in

Table 15.23 Common function to manage slave component references

```
type record of SyncSlaveSet SyncSlave;

const SyncSlaveSet c_emptySyncSlaveSet := {};

function f_addSyncSlaveSet ( in SyncSlave p_slave,
                             inout SyncSlaveSet p_set ) {
  p_set[ sizeof( p_set )] := p_slave;
  return;
}
```

Table 15.24 General synchronisation component type for synchronisation master

```
type component SyncMaster {
  port SyncMasterPort pt_sync;
  var SyncSlaveSet v_slaveSet := {}  // for managing slave component refs
};
```

Table 15.25 Common master synchronisation function to start a phase on slave components

```
function f_startPhase( in Phase p_phase ) runs on SyncMaster {
  var integer v_i;
  var integer v_amount := sizeof( v_slaveSet );
  var PhaseStartReq v_phaseStartReq := { phase := p_phase };

  for( v_i := 0; v_i < v_amount ; v_i := v_i + 1 ) {
    pt_sync.send( v_phaseStartReq ) to v_slaveSet[ v_i ]
  }
}
```

Table 15.25, two integer variables are defined. The variable v_i is a normal index variable while v_amount holds the number of PTCs. The third variable v_phaseStartReq holds the actual message to be sent. In the for-loop, the message is sent via the port pt_sync to all PTCs by using their references in v_slaveSet to indicate the recipient of the message.

To wait for all PTCs to finish a given phase, the MTC simply waits for the reception of as many messages of type PhaseEndInd as there are elements in v_slaveSet, see Table 15.26. In the first alternative we handle the consumption of the expected synchronisation phase. The second alternative is similar to the first one, except that the receive statement matches all messages, for example if a test component sends a synchronisation message pertaining to an earlier or later phase, and works as a 'catch all'. As these are non-expected messages, the verdict is set to inconc. A timer guards against the event that no message is received at all, for example a PTC is stuck in its execution. Please note that the handling of unexpected events can be handled in many different ways and we only present here one suggestion.

Table 15.26 Common master synchronisation function to wait for the end of a given phase from all slaves

```
function f_awaitEndPhase ( in template Phase p_phase ) runs on SyncMaster {
  var integer v_amount := sizeof( v_slaveSet );
  var integer v_i;

  for( v_i := 0; v_i < v_amount ; v_i := v_i + 1 ) {
    alt {
      [] pt_sync.receive( PhaseEndInd: { phase := p_phase } ) { }
      [] pt_sync.receive {
            setverdict( inconc ); // handling of out-of-sync

        }
      }
    }
}
```

The actual definition of the test case behaviour is rather schematic; the code is shown in Table 15.27. First, the PTCs are created one after the other. The references are added to the set of synchronisation slaves, the ports used in synchronisation are connected, and the behaviour on the newly created test component is started. After creating all the PTCs and starting their behaviour, the PTCs wait for the message to start the preamble phase. For each of the three phases, the MTC commands the PTCs to execute the behaviour for a given phase and thereafter waits until all test components finish that phase. Finally,

Table 15.27 Example use of master synchronisation functions

```
testcase tc () runs on MTC {
  var PTC1 v_ptc1;
  // set up test configuration
  v_ptc1 := PTC1.create;
  f_addSyncSlaveSet( v_ptc1, v_slaveSet );
  connect( mtc:pt_sync, v_ptc1:pt_sync );
  v_ptc1.start( f_ptc1() );
  // first   synchronisation point
  f_startPhase( e_preamble );
  f_awaitEndPhase( e_preamble );
  // second   synchronisation point
  f_startPhase( e_testbody );
  f_awaitEndPhase( e_testbody );
  // third   synchronisation point
  f_startPhase( e_postamble );
  f_awaitEndPhase( e_postamble );
    disconnect( mtc:pt_sync, v_ptc1:pt_sync );
  // ...
  stop;
}
```

after executing all phases, the ports for synchronisation are disconnected and the test case terminates. No additional synchronisation is needed to await termination of the PTCs. The PTCs stop themselves after they indicate that they have finished their postamble behaviour.

The code listed in this section is rather simplistic and probably should be further extended, re-factored and improved for use with real TTCN-3 test systems. The addition of timers and handling of other potential incorrect test component or SUT behaviour is needed in order to end up with a truly robust solution. Even with this in mind, the code provides a foundation, which is a good starting point that can be adapted to many different situations.

16

LTE Testing with TTCN-3[1]

In this chapter we will consider Long Term Evolution (LTE)[2] testing using TTCN-3. Associated with this book is a website which contains a working TTCN-3 tool and a version of the LTE test suite developed by ETSI for 3rd Generation Partnership Project (3GPP).[3] The web site URL is: www.wiley.com/go/willcock_TTCN-3_2e. Anyone wishing to access the material will need to complete a registration form online by following the link. Once the submitted registration form has been approved, you will receive a password for this site to enable the material to be downloaded.

16.1 LTE Description

LTE [1] is the latest standardized mobile telecommunications system which follows on from Universal Mobile Telecommunication System (UMTS) [48]. In comparison to earlier technologies LTE offers higher data transfer speeds and lower latency for the user. LTE has been developed by the 3GPP and the first complete version of the standards were released in 2008 (release 8), with the first commercial systems being rolled out during 2010. As part of this standardisation effort a test suite has been developed to check the conformance of LTE mobile devices (UEs) to the standards. It is this test suite which we will be considering in this chapter.

LTE offers a more efficient air interface and also introduces a new flatter architecture. The overall LTE architecture, including the Radio Access Network (RAN) and core part is shown in Figure 16.1. The interface between RAN and Core network is called the

---

[1] TTCN-3 Extracts and a few phrases/diagrams/tables are reproduced with permission from © 2010. 3GPP (TM).
[2] LTE is a trade mark of ETSI.
[3] © 2010. 3GPP™ TSs, and TRs are the property of ARIB, ATIS, CCSA, ETSI, TTA, and TTC who jointly own the copyright in them. They are subject to further modifications and are therefore provided to you "as is" for information purposes only. Further use is strictly prohibited.

An Introduction to TTCN-3, Second Edition.
Colin Willcock, Thomas Deiß, Stephan Tobies, Stefan Keil, Federico Engler and Stephan Schulz.
© 2011 John Wiley & Sons, Ltd. Published 2011 by John Wiley & Sons, Ltd.

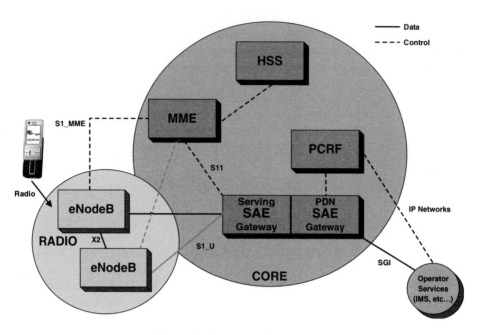

Figure 16.1   LTE architecture.

S1 interface and the interface between eNodeBs is named X2. The S1 interface has the following functionalities:

- It connects the eNode B to the evolved packet core. Divided into control plane (S1_MME) and user plane (S1_U) parts.
- S1_U carries the user data to SAE gateways.
- S1_MME connects to the mobility management entity.
- Carries the non access stratum (NAS) signalling (authentication and so on protocols between core and UE).

The X2 interface has the following functionalities:

- In inter- eNodeB handover to facilitate handover and provide data forwarding.
- In Radio Resource Management (RRM) to provide information like load levels to neighbouring eNodeBs to facilitate interference management.

16.2   LTE Test Suite

16.2.1   Test System Overview

The conceptual test architecture can be seen in Figure 16.2. The system under test (SUT) is the LTE mobile device (UE). The test system has two main components, the host and the hardware adaptor. The host element is where the test control, TTCN-3 execution engine and associated codecs reside. The hardware adaptor provides the adaption to the SUT. In

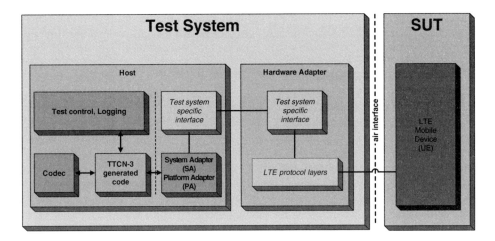

Figure 16.2 LTE test system architecture.

this case this means hosting all the necessary lower protocol layers below the TTCN-3 test level plus the hardware interface to the UE.

16.2.2 LTE Test Suite Overview

Once you have the LTE test suite in the TTCN-3 tool and start looking through the code, you will notice this is not a simple test suite. Indeed it is too complex to cover all the details in this book, instead we will try to provide a first overview which will help you understand the overall structure and testing ideas.

While considering the overall test suite you may also notice that it incorporates many of the ideas and suggestions we have seen in Chapters 14 and 15, the test suite uses a well defined test framework. It has a well defined naming convention. For example, functions running on a component start with f_, altsteps start with a_, templates for sending start with cs_ and templates for receiving start with cr_. The full list of naming conventions can be seen from Table B.3.1 in [44]. As well as a clear naming convention the test suite also defines a clear layout for the declarations, for example all local variables are declared at the beginning of a function and the order of declarations is local constants followed by local variables and then local timers.

At the uppermost level the test suite is divided into a series of folders with the prefix ttcn3. Within each of these folders is one or more TTCN-3 modules. Let us start by looking at the first folder which has the name ttcn3. When we double click on this folder to see its contents we can see that it contains two TTCN-3 modules, LTE_EPS_TS_SelectionExpressions and LTE_EPS_TS_Testcases. If we now double click on the module LTE_EPS_TS_Testcases the source code of this module should be displayed in the middle window. This module can be seen as the central module for the whole test suite, it defines all the test cases. If we look a bit closer at the contents we see that it starts with a series of import statements pulling in definitions from the other modules, then it defines all the test cases and lastly we have the control part (see Section 3.2.6) where the test cases are actually called. Whether a specific test case is actually called when the control part is executed depends on the result from the function f_ExecutionGuideline in some cases combined with test case selection expressions.

16.2.3 Test Case Definitions

Let us now take a look at one specific test case to get a better idea about how things are structured and how the test component architecture is organised. If we look at test case TC_6_1_1_1 shown in Table 16.1, we can see that the test case runs on the master test component MTC with the system definition SYSTEM (see Section 5.2.1). After the comment defining the purpose we then see three variables declared which provide references to possible parallel test components. Through the naming it appears clear that we have one parallel test component for each possible radio access technology which we might wish to use in this test suite. GERAN_PTC for GSM, UTRAN_PTC for UMTS and EUTRA_PTC for LTE (even though this test suite is testing LTE there are certain tests which require a radio link to be created between the test system and the SUT using one of these other radio access technologies, for example for handover testing). In the case of our chosen test case we only actually use the LTE connection therefore it is only the LTE PTC that we actually instantiate with the create command (see Section 5.2.3). After creating the necessary parallel test components the test case defines a guard timer and then calls the function f_MTC_ConnectPTCs. This function makes all the necessary connections between the MTC and the PTCs. After creating and connecting the necessary components we are now at the point to actually start the required test behaviour. This is done using the start operation on the LTE PTC passing in the function f_TC_6_1_1_1_EUTRA. It is in this function that the actual test case behaviour, sending and receiving messages with the SUT, is defined. We shall return to this function later. Next the test case starts the guard timer and calls the function f_MTC_MainLoop which basically just waits for parallel test components to complete.

From this first look at the test case TC_6_1_1_1 we can make a number of observations and indeed if you look at all the other test cases in this module you will see that these observations are true for the whole test suite.

Table 16.1 Example test case from LTE test suite

```
testcase TC_6_1_1_1() runs on MTC system SYSTEM {
    // @purpose
    //    PLMN selection of RPLMN, HPLMN/EHPLMN, UPLMN and OPLMN
    var GERAN_PTC    v_GERAN      := null;
    var EUTRA_PTC    v_EUTRA      := EUTRA_PTC.create alive;
    var UTRAN_PTC    v_UTRAN      := null;
    timer t_GuardTimer := int2float(2400);

    f_MTC_ConnectPTCs(v_EUTRA, v_UTRAN, v_GERAN);
    v_EUTRA.start    (f_TC_6_1_1_1_EUTRA());
    t_GuardTimer.start;
    f_MTC_MainLoop(t_GuardTimer);
}
```

Firstly we can see that the top level test cases are executed exclusively on the MTC. Secondly we can see that the actual test behaviour of the test cases is executed on the parallel test components. The test cases running on the MTC all have the same tasks: create the PTCs, connect the PTCs, start the PTCs, start the guard timer and then wait for PTCs to finish.

16.2.4   Test Behaviour Definition

Now it is time to look at the actual test behaviour definition for this test case which is defined in the function f_TC_6_1_1_1_EUTRA. The definition for this function can be found in the module Idle_PLMNSelection in the folder ttcn3\6_1. The behaviour in the function can be divided into three parts. The first part is the preamble, getting the SUT in the required state to perform the test, then comes the test body, the actual behaviour we are trying to test and lastly comes the postamble to return the SUT to known state to enable further test cases to be run in a controlled manner. The easiest way to locate where these various parts start and finish is the calls to the f_EUTRA_TestBody_Set function. This function is called with parameter true at the beginning of the test body and called again with parameter false as the end of the test body. The test body is itself divided into a series of steps. These steps correspond to the test description in [49].

16.2.5   EUTRA Parallel Test Component

Now let us consider the parallel test component that this function is designed to run on. As explained in Section 4.10 the component type for such a function must be given in the runs on clause. From f_TC_6_1_1_1_EUTRA we can see that the component type is EUTRA_PTC. This type is defined in the module EUTRA_Component in the folder ttcn3/commonEUTRA. The definition for this type is also shown in Table 16.2 and shown visually in Figure 16.3. In general the ports can be divided into four groups. The first shown at the top of Figure 16.3 is the interface to the MTC. As we have seen this UT interface is used to control the execution of the behaviour on the parallel test component. The second group of ports on the right-hand side of the diagram are the coordination and communication interfaces to the other parallel test components. On the left-hand side of the diagram are the ports that interface with the protocol stack and emulators that sit below the TTCN-3 execution engine. Lastly at the bottom of the diagram is the set of ports that communicate with the SUT. The SRB port is used for control plane messages and the DRB port is used for user plane data. The control plane is the message path used for the signalling traffic between the LTE network and mobile device. The user plane is the message path for all user data, that is the actual messages that the user of the mobile device wants to send or receive.

16.2.6   Test Suite Module Structure

Having considered the test case TC_6_1_1_1 and the associated function f_TC_6_1_1_1_EUTRA we are now in a position to better understand the overall module structure for the test suite. We can see that the modules containing the test behaviour running on the

Table 16.2   Definition of EUTRA PTC

```
type component EUTRA_PTC {
    var   EUTRA_Global_Type vc_EUTRA_Global;

    port EUTRA_SYSTEM_PORT        SYS;
    port EUTRA_SYSIND_PORT        SYSIND;
    port EUTRA_SRB_PORT           SRB;
    port EUTRA_NASCTRL_PORT       NASCTRL;
    port EUTRA_DRB_PORT           DRB;

    port UT_PTC_MTC_PORT          UT;

    port IRAT_CO_ORD_PORT         UTRAN;
    port IRAT_CO_ORD_PORT         GERAN;

    port IP_RAT_CTRL_PORT         IP
};
```

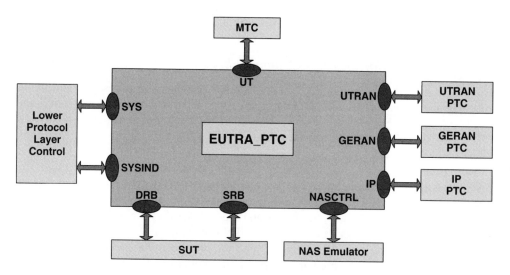

Figure 16.3   EUTRA component type.

parallel test components are grouped in folders reflecting the structure and numbering from the conformance specification part [49]. These folders all have the name form `ttcn3\n` or `ttcn3\n_n` where `n` is a number. The areas of testing covered by these groups are shown in Table 16.3.

Around 50% of the folders contain these testing group modules, the other folders contain modules which cover the common parts of the tests. These common modules contain common parts for particular radio access technologies or for specific types of testing, like layer 2 tests. The folders names have the word Common in their names.

Table 16.3 Test group areas for LTE test suite

Group	Sub-group	Test area
6	–	Idle mode operations
	6.1	In a pure E-UTRAN environment
	6.2	Multi-mode environment (E-UTRAN, UTRAN, GERAN, CDMA2000)
	6.3	Closed subscriber group cells
7	–	Layer 2
	7.1	MAC
	7.2	RLC
	7.3	PDCP
8	–	RRC
	8.1	RRC connection management procedures
	8.2	RRC connection reconfiguration
	8.3	Measurement configuration control and reporting
	8.4	Inter-RAT handover
	8.5	RRC others
9	–	EPS mobility management
	9.1	EMM common procedures
	9.2	EMM specific procedures
	9.3	EMM connection management procedures (S1 mode only)
	9.4	NAS Security
10	–	EPS session management
11	–	General tests
12	–	E-UTRA radio bearer tests
13	–	Multi layer procedures

The common parts are typically reused in many of the testing functions defined in the testing group modules. This structuring of modules follows the principles of frameworks described in Chapter 14.

16.2.7 RRC Message Definitions

The most important protocol for this test suite is the Radio Resource Control (RRC) Protocol [50]. This is the layer 3 LTE protocol which is used by most of the TTCN-3 test cases to exchange signalling messages with the SUT. The RRC protocol is defined using ASN.1 [39] and as described in Section 9.5.1 the types are imported and directly used within the TTCN-3 definitions. Returning to module `Idle_PLMNSelection` in the folder `ttcn3\6_1` we can see at the beginning of the imports section:

```
import from EUTRA_RRC_ASN1_Definitions language "ASN.1:1997" all;
```

This definition pulls in the RRC message type definitions. Using the TTCN-3 tool we can directly look at these ASN.1 definitions. The ASN.1 module can be found in the folder `ttcn3\CommonEUTRA_Defs` with the name `EUTRA_RRC_ASN1_Definitions.asn`. You will notice that there is another module here with the same name but a different file extension `ttcn3view`. This module enables you to see these ASN.1 definitions translated into the TTCN-3 syntax. This might make it easier for you to follow.

Now in the final part of this chapter let's take a look at how these ASN.1 type definitions are used in the test cases. Let us return to the function `f_TC_6_1_1_1_EUTRA` defined in the module `Idle_PLMNSelection` in the folder `ttcn3\6_1` and consider step 3 in the test body as shown in Table 16.4.

We can see that this step starts with a receive statement. From our earlier consideration of the `EUTRA_PTC` component type we know that because the receive statement is on the SRB port it is expecting a signalling message from the SUT (the LTE mobile device). The expected message is defined by the template `cas_SRB0_RrcPdu_REQ`. If we look at the definition of this template, which is in the module `EUTRA_SRB_Templates` in the folder `ttcn3\commonEUTRA_Templates` and shown in Table 16.5, we can see that the actual type that we are expecting to receive is `SRB_COMMON_REQ`. We can also note that the template and the parameters to the template make use of the template restrictions described in Section 11.6. The `(value)` notation specifies that the templates must resolve to a specific value.

The type definition for `SRB_COMMON_REQ` can be found in the module `EUTRA_ASP_Srbdefs` in the folder `ttcn3\CommonEUTRA_Defs` and in Table 16.6. From the type definition we can see that the message is divided into two parts the Abstract Service Primitive (ASP) information and the signalling. In this test suite all the messages are divided into an ASP and Protocol Data Unit (PDU). The actual information that the

Table 16.4 Example test step

```
//     Step 3: Receive RRCConnectionRequest on Cell 12
SRB.receive(car_SRB0_RrcPdu_IND(eutra_Cell12,
                       cr_508_RRCConnectionRequest));
t_IdleMode_GenericTimer.stop;
//* @verdict pass RRCConnectionRequest message received on Cell 1
f_EUTRA_PreliminaryPass(__FILE__, __LINE__, "Test Case 6.1.1.1 Step 3");
```

Table 16.5 Example SRB template

```
template (value) SRB_COMMON_REQ cas_SRB0_RrcPdu_REQ(
                                        CellId_Type p_CellId,
                         template (value) TimingInfo_Type p_TimingInfo,
                         template (value) DL_CCCH_Message p_RrcPdu) :=
Common := cs_ReqAspCommonPart_SRB(p_CellId, tsc_SRB0, p_TimingInfo),
Signalling := {
  Rrc := {
    Ccch := p_RrcPdu
  },
  Nas := omit
  }
};
```

Table 16.6 SRB_COMMON_REQ type definition

```
type record SRB_COMMON_REQ {
/* common ASP to send PDUs to SRB0, SRB1 or SRB2 */
    ReqAspCommonPart_Type         Common,
    C_Plane_Request_Type          Signalling
};
```

Table 16.7 C_Plane_Request_Type definition

```
type record C_Plane_Request_Type {
   /* RRC and/or NAS PDU to be send to the UE */

   RRC_MSG_Request_Type         Rrc          optional,
   NAS_MSG_RequestList_Type     Nas          optional
};
```

Table 16.8 RRC_MSG_Request_Type definition

```
type union RRC_MSG_Request_Type {
/* DL RRC PDU on CCCH or DCCH */
    DL_CCCH_Message         Ccch,
    DL_DCCH_Message         Dcch
  };
```

SUT will receive is defined in the PDU, in this example contained in the field `Sig-nalling`. The ASP and any associated information fields associated with it are purely for the underlying protocol stacks and adaption layers that sit below the TTCN-3 test execution engine, see Figure 16.2.

If we want to find the point where the ASN.1 definitions are used we need to take a closer look at the PDU part of this type which in our case should resolve to a RRC message. The type of the Signalling field which contains the PDU is `C_Plane_Request_Type`. Its type definition can be found in the same module as `SRB_COMMON_REQ` and in Table 16.7. In this test case, which involves the RRC protocol it is only the first field `Rrc`, that is relevant. This field has the type `RRC_MSG_Request_Type` which can be found in the module `EUTRA_CommonDefs` in the folder `ttcn3\CommonEUTRA_Defs` and in Table 16.8. Looking at this type we finally have the link to the ASN.1 definitions. If we look at the first field in this type which is the relevant one for us the field `Ccch` is of type `DL_CCCH_Message`. This type is defined in the `EUTRA_RRC_ASN.1_Definitions` module we considered earlier.

16.3 Summary

This chapter has clarified that LTE is the latest standardized mobile network system which is beginning to be deployed, offering advantages in terms of data throughput and latency.

We briefly covered the basic architecture based on a new flat hierarchy and explained as part of the standardization efforts at 3GPP a TTCN-3 test suite has been developed to test LTE mobile devices. In the rest of the chapter we took a first look at this test suite, looking at one particular test case to understand the overall structure and test system architecture.

The 3GPP efforts to provide a complete standardized conformance test suite for LTE mobile devices is still on-going and due to the continued new features and frequencies being added to the LTE standards with each new release, there is no end in sight for this development and expansion of the LTE test suite. The latest version of the test suite can be downloaded from: http://www.3gpp.org/ftp/tsg_ran/WG5_Test_ex-T1/TTCN/Deliveries/ LTE_SAE/.

17

Closing Thoughts and Future Directions

TTCN-3 is a living language. As these words are written, there is active work going on at the European Telecommunication Standardisation Institute (ETSI) to maintain and extend the language. Lately, in addition to extending the core notation, several extension packages have been defined which we have also outlined in this book. Extension packages introduce, in a more self contained manner, new language concepts for advanced uses and new application areas of TTCN-3 such as performance and real-time testing.

A great source of information is ETSI's official TTCN-3 website `http://www.ttcn-3.org`. Here, you can download the latest version of the TTCN-3 language standards documents and find out how to join the TTCN-3 mailing list, an open forum to ask questions and discuss issues about the use and development of the TTCN-3 language. In addition, you can find links to all of the known TTCN-3 tools, open source developments, public test suites, tutorials, presentations from all the TTCN-3 User Conferences, as well as information about how to become a certified TTCN-3 test engineer. Finally, it also provides instructions for submitting change requests on any of TTCN-3 standards or the extensions. This mechanism can be used by anyone to report errors in the standard, defects or inconsistencies in the language, or to propose new extensions. Change requests are reviewed regularly by ETSI's TTCN-3 maintenance group.

As we write this book, it appears, at least in the telecommunications domain, that we face a number of new testing challenges. One of these challenges is the increasing complexity of the products that we must develop, combined with the pressure to shorten the time to market and improve quality. This leads to questions of how we can develop complex systems faster and better and therefore logically how we can improve testing by automating test execution. In general, key goals are to reduce test execution time and making test execution more flexible as well as repeatable. However, automation alone is not a warrant for improving testing – tests have to be specified in the correct way. Based on our experience of applying TTCN-3 in real industrial cases, we believe that this standardised testing language is an excellent basis for writing high quality tests.

An Introduction to TTCN-3, Second Edition.
Colin Willcock, Thomas Deiß, Stephan Tobies, Stefan Keil, Federico Engler and Stephan Schulz.
© 2011 John Wiley & Sons, Ltd. Published 2011 by John Wiley & Sons, Ltd.

Remember that at its roots TTCN already today embodies the expertise and experience from 20 years of testing which differentiates it from any other scripting language.

Many of these testing challenges are not unique to the telecommunication industry – but the solution that TTCN-3 provides are generally applicable. Proof of this is the rapid adoption of TTCN-3 since the writing of our first edition of this book: TTCN-3 is now well established in the telecommunication sector – in addition to LTE TTCN-3 has also been heavily used with other technologies such as the IP Multimedia System (IMS), [52] WiMax [4] and TETRA [53]. Convergence of telecom and Internet has brought TTCN-3 to Internet protocols like SIP [26] and IPv6 [45]. TTCN has been adopted by the automotive sector and massive efforts are on the way in the related sector of intelligent transport systems. In addition, TTCN-3 has managed to spread from pure functional conformance testing into interoperability testing and load testing. But these are just examples from the standardisation domain. We should not overlook that TTCN-3 has also silently managed to be used as a general purpose test specification language in product development.

Looking at the future development and use of TTCN-3 in the short term, we would expect the language to continue to replace proprietary test languages and existing standardised test languages in functional testing, and stay on its path to become the unifying testing technology across application domains and different testing types. Starting from the recently defined extension packages, TTCN-3 is likely to move further into real time testing and open doors for new application areas such as industrial automation.

New approaches to testing are pushing to the market such as model-based testing. Here, test scripts, for example in TTCN-3, are automatically generated from abstract models. Model-based testing has been shown to enable significant improvements in productivity of test engineers and test suite quality but it is anticipated that it will not be able to eliminate the need for hand written segments and certain types of tests. ETSI has started recently to define how model-based testing relates to TTCN-3 test specification. In the medium term, we expect that integration of model-based testing in test suite development will become much better defined and start to establish itself at least in functional testing.

In the long term, a new extension for continuous systems could enable the use of TTCN-3 in even more application areas. A number of research projects and initiatives have already explored this area of application but it still remains to be seen to what extent industry and ETSI will see a need and use for such an extension.

At the very beginning, when this book was just at the initial concept phase, we imagined translating our experience in TTCN-3 language standardisation, tool development, and industrial use into a practical guide for those who wish to get the most from this powerful testing technology. When writing this second edition of the book, we not only updated the book to the latest approved edition of the standard, but also tried to enhance the book with additional experience gained in several standardisation and industrial projects with partners from academia and industry. If you have had the patience and understanding to read this far, we hope you feel we have at least to some extent managed to meet that initial goal. In the final analysis, whatever the hopes and goals of the authors, the success or failure of a book is decided by its readers.

References

[1] Holma, H. Dr. and Toskala, A. Dr. (2009) *LTE for UMTS – OFDMA and SC-FDMA Based Radio Access*, John Wiley & Sons, Inc.

[2] http://www.3gpp.org/About-3GPP.

[3] http://www.openmobilealliance.org/AboutOMA/Default.aspx.

[4] http://www.wimaxforum.org/about.

[5] http://www.autosar.org/index.php?p=1&up=0&uup=0&uuup=0.

[6] http://ipv6ready.org/?page=about.

[7] http://www.etsi.org/WebSite/Standards/Interoperability.aspx.

[8] ETSI ES 201 873-1. *Methods for Testing and Specification (MTS), The Testing and Test Control Notation Version 3; TTCN-3 Core Language*.

[9] ETSI ES 201 873-2. *Methods for Testing and Specification (MTS), The Testing and Test Control Notation Version 3; Tabular Presentation Format*.

[10] ETSI ES 201 873-3. *Methods for Testing and Specification (MTS), The Testing and Test Control Notation Version 3; Graphical Presentation Format*.

[11] ETSI ES 201 873-4. *Methods for Testing and Specification (MTS), The Testing and Test Control Notation Version 3; Operational Semantics*.

[12] ETSI ES 201 873-5. *Methods for Testing and Specification (MTS), The Testing and Test Control Notation Version 3; Runtime Interface*.

[13] ETSI ES 201 873-6. *Methods for Testing and Specification (MTS), The Testing and Test Control Notation Version 3; ES 201 873-6 TTCN-3 Control Interface*.

[14] ETSI ES 201 873-8. *Methods for Testing and Specification (MTS), The Testing and Test Control Notation Version 3; Part 8: The IDL to TTCN-3 Mapping*.

[15] ETSI ES 201 873-9. *Methods for Testing and Specification (MTS), The Testing and Test Control Notation Version 3; Part 9: Using XML Schema with TTCN-3*.

[16] ETSI ES 201 873-10. *Methods for Testing and Specification (MTS), The Testing and Test Control Notation Version 3; Part 10: TTCN-3 Documentation Comment Specification*.

[17] ETSI ES 202 785. *Methods for Testing and Specification (MTS), The Testing and Test Control Notation Version 3; TTCN-3 Language Extensions: Behaviour Types*.

[18] ETSI ES 202 784. *Methods for Testing and Specification (MTS), The Testing and Test Control Notation Version 3; TTCN-3 Language Extensions: Advanced Parameterization*.

[19] ETSI ES 202 781. *Methods for Testing and Specification (MTS); Testing and Test Control Notation Version 3 Extension; Package Configuration and Deployment Support*.

[20] ETSI ES 201 873-7. *Methods for Testing and Specification (MTS), The Testing and Test Control Notation Version 3; Part 7: Using ASN.1 with TTCN-3*.

[21] ETSI ES 202 782. *Methods for Testing and Specification (MTS); Testing and Test Control Notation Version 3 Extension; Performance and Real Time Testing*.

[22] ITU-T Recommendation Z.120. *Message Sequence Chart*.

An Introduction to TTCN-3, Second Edition.
Colin Willcock, Thomas Deiß, Stephan Tobies, Stefan Keil, Federico Engler and Stephan Schulz.
© 2011 John Wiley & Sons, Ltd. Published 2011 by John Wiley & Sons, Ltd.

[23] OSI 7498. *OSI 7 Layer Model*.

[24] ITU-T Recommendation X.292. *Tree and Tabular Combined Notation*.

[25] ISO/IEC 9646-3. *Tree and Tabular Combined Notation*.

[26] IETF RFC 3261. *Session Initiation Protocol*.

[27] ITU-T Recommendation X.920. *Interface Definition Language*.

[28] W3C Recommendation. *Extensible Markup Language*.

[29] IETF RFC 1035. *Domain Names – Implementation and Specification*.

[30] IETF RFC 2616. *Hypertext Transfer Protocol – HTTP/1.1*.

[31] IETF RFC 2821. *Simple Mail Transfer Protocol*.

[32] IETF RFC 959. *File Transfer Protocol (FTP)*.

[33] ISO/IEC 10646. *Information Technology – Universal Multiple-Octet Coded Character Set (UCS)*.

[34] ISO/IEC 14750:1999. *Information Technology – Open Distributed Processing – Interface Definition Language*.

[35] Schulz, S. (2004) Derivation of abstract protocol type definitions for the conformance testing of text-based protocols. *Proceedings of 16th International Conference on Testing of Communicating Systems (TestCom), Oxford*, pp. 177–192.

[36] ISO/IEC 8652:1999. *Information Technology – Programming Languages – Ada*.

[37] Nelson, G. (ed.) (1991) *Systems Programming with Modula-3*, Prentice Hall.

[38] IEEE Standard 1076-2000. *VHDL Language Reference Manual*.

[39] ITU-T Recommendation X.680. *Information Technology – Abstract Syntax Notation One (ASN.1): Specification of Basic Notation*.

[40] ITU-T X.694 (2008-11) *Information Technology – ASN.1 Encoding Rules: Mapping W3C XML Schema Definitions into ASN.1*.

[41] 3GPP TS 29.002. *Mobile Application Part (MAP) Specification*.

[42] Schulz, S. and Vassiliou-Gioles, T. (2002) Implementation of TTCN-3 test systems using the TRI. *Proceedings of 14th International Conference on Testing of Communicating Systems (TestCom), Berlin, Germany, April 2002*, pp. 425–441.

[43] ETSI TR 102 788 V1.1.1 (2010-01) *Technical Report, Methods for Testing and Specification (MTS); Automated Interoperability Testing; Specific Architectures*.

[44] 3GPP TS 36.523-3 V8.3.0 (2010-06). *3rd Generation Partnership Project; Technical Specification Group Radio Access Network; Evolved Universal Terrestrial Radio Access (E-UTRA) and Evolved Packet Core (EPC); User Equipment (UE) Conformance Specification Part 3: Test Suites (Release 8)*.

[45] ETSI TS 102 351 V2.1.1 (2005-08) *Technical Specification, Methods for Testing and Specification (MTS); Internet Protocol Testing (IPT); IPv6 Testing: Methodology and Framework*.

[46] Dijkstra, E.W. (1968) Goto considered harmful. Communications of the ACM, **11** (3), 147–148.

[47] ISO/IEC 9646-1:1994. *Information Technology – Open Systems Interconnection – Conformance Testing Methodology and Framework – Part 1: General Concepts*.

[48] Holma, H. Dr. and Toskala, A. Dr. (2006) *HSDPA/HSUPA for UMTS*, John Wiley & Son, Inc.

[49] 3GPP TS 36.523-1 V8.3.0 (2010-06) *3rd Generation Partnership Project; Technical Specification Group Radio Access Network; Evolved Universal Terrestrial Radio Access (E-UTRA) and Evolved Packet Core (EPC); User Equipment (UE) Conformance Specification Part 1: Protocol Conformance Specification (Release 8)*.

[50] 3GPP TS 36.331 V9.3.0 (2010-06). *3rd Generation Partnership Project; Technical Specification Group Radio Access Network; Evolved Universal Terrestrial Radio Access (E-UTRA); Radio Resource Control (RRC); Protocol Specification (Release 9)*.

[51] ITU-T Recommendation X.660:1992. *Information Technology – Open Systems Interconnection – Procedures for the Operation of OSI Registration Authorities: General Procedures*.

[52] 3GPP TS 23.228. 3rd Generation Partnership Project; Technical Specification Group Services and System Aspects. IP Multimedia Subsystem (MS) Stage 2.

[53] ETSI EN 300 392-1; Terrestrial Trunked Radio (TETRA), Voice plus Data (V+D); Part 1 General Network Design.

Index

An Introduction to TTCN-3, Second Edition.
Colin Willcock, Thomas Deiß, Stephan Tobies, Stefan Keil, Federico Engler and Stephan Schulz.
© 2011 John Wiley & Sons, Ltd. Published 2011 by John Wiley & Sons, Ltd.